"I have never seen a synth a whole field performed as one area is developing. Your integration of aspects of molecular coding with principles of classical biology is nothing short of a virtuoso feat. I have learned a great deal from it. Your book will be a priceless contribution, coming as it does right in the midst of this rapidly developing field"—*Ernest Borek, author of THE CODE OF LIFE and THE ATOMS WITHIN US, in a letter to the author.*

A definitive statement of the chemical basis of evolution, this book describes in detail the present knowledge and success that science has acquired in deciphering the evolutionary history of living organisms.

of proteins and the genetic code. The author discusses the new knowledge of the chemical nature of mutations, of the base sequences in the genetic code, of the mechanism of protein synthesis, and of gene duplication.

This book inaugurates the Columbia Series in Molecular Biology, of which Ernest Borek is advisory editor.

MOLECULES AND EVOLUTION

MOLECULES AND EVOLUTION

Thomas H. Jukes

Very old are the rocks.
The pattern of life is not in their veins.
   When the earth cooled, the great rains
   came and the seas were filled.
   Slowly the molecules enmeshed in
   ordered asymmetry.
   A billion years passed, aeons of
   trial and error.
   The life message took form, a spiral,
   a helix, repeating itself endlessly,
   Swathed in protein, nurtured by
   enzymes, sheltered in membranes,
   laved by salt water, armored with
   lime.
Shells glisten by the ocean marge,
Surf boils, sea mews cry, and the great wind
soughs in the cypress.

Columbia University Press   New York and London   1966

Thomas H. Jukes is Professor in Residence and heads the biochemical and physiological programs in the Space Sciences Laboratory at the University of California at Berkeley. He was formerly director of a research section at Lederle Laboratories Division, American Cyanamid Co.

The author's research was supported by Grant NsG 479 from the National Aeronautics and Space Administration and by grants from the National Institutes of Health to the University of California.

The quotation on p. 253 is from *General Zoology* by T. I. Storer and R. L. Usinger. Copyright © 1965 McGraw-Hill Book Company, Inc. Used by permission of McGraw-Hill Book Company.

Frontispiece: "Rocks and Limpets, Point Lobos, 1963." Photograph by Ansel Adams. From an enlargement of a Polaroid Land 4 × 5 Type 55 P.N. Negative.

Copyright © 1966 Columbia University Press
Library of Congress Catalog Card Number: 66-19974
Printed in the United States of America

Preface

Life may be regarded as a step in the evolutionary history of our planet. Organisms owe their life to a process that transmits information from generation to generation. Reproduction of mutational changes takes place in this process; the persistence of such changes is subject to competitive pressures in the "struggle for existence." Nonliving mixtures of chemicals do not currently give rise to life, but it is necessary to postulate that such a step must have occurred at some time in the past. A common pattern in chemistry of life is in contrast with the extreme diversity of living organisms; this paradox is, however, concordant with the viewpoint that these have evolved in numerous directions from a single origin.

Some of the most eloquent and monumental writings in science have been on the subject of evolution, and it is with some trepidation that I have ventured to address myself to an aspect of this massive topic.

I have gained some reassurance by imagining the wonder and delight that Charles Darwin and Thomas Huxley would have shown if they had learned that thyroid extract will produce metamorphosis from the perpetual larval stage in certain salamanders or that fertilization of the egg of an animal is followed by activation of the protein-synthesizing system or if they had read of the experiment by Weigert and Garen in which a drop of a mutagenic chemical produced seven different changes in a coding triplet when added to a bacterial culture, each change favoring survival of the bacteria by restoring the function of a defective gene.

Most books on evolution have dealt with entire organisms in terms of their structure, their separation into species, and their behavior as members of a population group. Within the past decade, a new interest has arisen; it centers on the implications of deoxyribonucleic acid

Preface

(DNA) as the carrier of genetic information. The continuity of terrestrial life depends upon the passage of molecules of DNA from generation to generation; the processes of evolution have been carried out as a result of changes in these molecules. This book reviews some of the research dealing with biochemical reactions that are responsible for the structure, function, and survival of living organisms and that have a bearing on evolution.

The susceptibility of biochemistry to evolutionary exploration is shown by comparing the amino acid sequences in proteins of different organisms. Such comparisons lead to postulations concerning the length of time that has been occupied by terrestrial life, and these lead in turn to speculations concerning the possible existence of life elsewhere in the solar system. An exploration of the surface of Mars is a proposal that has been widely discussed. Among the objectives would be the determination of the physical and chemical conditions of the Martian surface as a potential environment for life, the determination of whether or not life is or has been present on Mars, the characteristics of living organisms, if present, and, if none are present, the pattern of chemical evolution without life.

I wish to thank Ernest Borek and Robert Tilley, who instigated and encouraged the writing of the book, and my colleagues Harold Kammen, Joseph Krakow, Hiroshi Matsubara, and Hiroshi Yoshikawa, who read portions of the manuscript and drew my attention to errors (any uncorrected errors are, as the customary statement points out, the responsibility of the author). I wish also to thank Dorothy Walker and Dana Breaux, who typed the manuscript, Ellis Myers, who prepared most of the illustrations, Ansel Adams, who lent his photograph of "Rocks and Limpets," and Katherine M. Purcell, who edited the manuscript. I also thank the National Aeronautics and Space Administration for support by a grant, NsG 479, to the University of California, Berkeley, for a study of the chemistry of living systems, and Orr Reynolds for his interest and encouragement.

THOMAS H. JUKES

Berkeley, California
October, 1965

Contents

	Preface	v
1.	Evolution and Deoxyribonucleic Acid	1
2.	The Genetic Code	11
3.	Microbiology and the Study of Heredity	75
4.	Mutations	100
5.	Evolution and the Hemoglobins	146
6.	The Cytochromes c and Other Proteins Showing Evolutionary Changes	191
7.	Homology and Deoxyribonucleic Acid	230
8.	Biochemistry and Evolutionary Pathways	245
	Summary	265
	Man	266
	References	267
	Index	283

MOLECULES AND EVOLUTION

[1]

Evolution and Deoxyribonucleic Acid

> Nihil est toto, quod perstet, in orbe
> Cuncta fluunt, omnisque vagans formatur imago
> Ipsa quoque odsiduo labuntur tempora motu.
>
> There is nothing in the whole world which is permanent
> Everything flows onward; all things are brought into
>   being with a changing nature;
> The ages themselves glide by in constant movement.
>
> Ovid, *Metamorphoses* xv, i.177

Many of the boundaries that divide the various fields of biology are rapidly disappearing because of the advance of biochemical knowledge. It has long been known that resemblances exist between numerous types of living organisms. As the biologist traces back animal and plant life to the simplest forms, he finds single-celled organisms that resemble members of the animal kingdom yet contain chlorophyll, the pigment of green plants. Biochemists have discovered that in many cases the same intricate molecules are present in animals, flowering plants, and bacteria. Even the proteins that are largely responsible for the differences among all living organisms are synthesized from an identical set of twenty amino acids. The family tree of evolution shows that most forms of life have gradually diverged from each other through a long series of small changes. On every hand we see evidence that the complexities which typify life are built by combining small units into a seemingly limitless number of different patterns.

The concept of evolution extends into various fields of study. Broadly speaking, evolution is the product of the interaction of change and time. Its progress has been noted and examined by cosmologists,

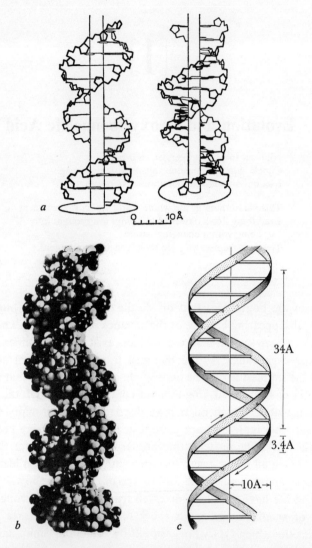

Fig. 1-1. (*a*) Diagrams of the Watson-Crick structure of DNA. The spiral lines represent the phosphate ester chains and the "rungs of the ladder" indicate the base pairs (A with T and G with C) joined by hydrogen bonds. (*b*) A diagram to illustrate the dimensions of the helix. The opposite polarity of the two strands is shown by arrows. (*c*) A molecular model of the helix. (From Borek, 1965, p. 167.)

philologists, and biologists. Evolution moves incessantly forward: new forms emerge that are adapted to their surroundings; their outmoded predecessors disappear from the scene. Biological evolution is a vast and intricate subject, but the principles that govern it are not haphazard. We shall discuss certain biochemical phenomena that underlie the evolutionary processes in living creatures.

The mechanism that controls life rests in a substance that carries information from generation to generation. This substance is deoxyribonucleic acid (DNA). The general nature of the structure of the DNA molecule has been frequently described and widely publicized; an essential feature is the complementarity between the base pairs adenine-thymine and guanine-cytosine, as shown in Fig. 1-1.

The complementarity is sometimes termed *Watson-Crick pairing* in allusion to its discoverers (Watson and Crick, 1953a), who first pointed out that the four bases—adenine, cytosine, guanine, and thymine (A, C, G, and T)—in DNA had molecular configurations that led to hydrogen bonding between A and T and between G and C in the two parallel strands of the DNA molecule. The pairing is duplicated when the two strands are separated and resynthesized enzymatically in the living cell, so that the original DNA parent molecule gives rise to two identical daughter molecules, each of which contains one parental strand.

The base sequences in the DNA molecules of each individual carry not only the information that characterizes the phenotype but also a cryptogrammic documentation of the evolutionary development of the individual. The continuity of the molecules reaches back to the origin of life on earth. The elementary beginnings of the translation of small fragments of such cryptograms are now appearing in the form of the amino acid sequences of a few characteristic proteins obtained from a handful of different species.

The method by which information necessary for the perpetuation of life is carried from generation to generation could well have been so complex as to be beyond our grasp. The discovery of the molecular structure of DNA showed that it was unnecessary to assume that any such complexity existed. Instead, a fine example was revealed of the intricacy that can be developed by Nature through the utilization of two simple principles, linearity and polarity, which are essentially the same basic properties that are found in a language. It is easy to learn the alphabet, yet who can read all the books that have been written with it? But who would shrink from the task of reading a book merely because there are many more volumes in the library than he can hope to comprehend?

The "alphabet" of DNA is, of course, the four bases adenine, cytosine, guanine, and thymine. A molecule of DNA can contain as many as 10 million of these four "letters" arranged in two parallel rows, which are "read" in opposite directions. Why there should be two rows instead of one is not immediately obvious, but several suggestions come to mind. It may be that a double row is necessary for the replicating mechanism in DNA, for the production of a single-strand of messenger ribonucleic RNA which transmits the inherited information to the protein-synthesizing process, or for the protection of DNA against some of the enzymes that attack single-stranded nucleic acids. It is also possible that the double strand protects the molecule to some extent against the effects of changes in one of its strands. A double strand can hold itself together by hydrogen bonding if a break or the removal of a few bases occurs in one of its two components. The damage in the single strand can be repaired enzymatically.

A sequence of 64 letters using A, C, G, and T in all possible permutations can be written in 4^{63} different ways. This number is even more impressive than examples of large numbers taken from astronomy; it is about 2 billion billion billion billion. Such a sequence, in terms of DNA, would represent only about one-seventh of the length of a typical gene. The sequence of bases in a haploid mammalian cell is much longer, about 3 billion bases. This sequence can theoretically be arranged in a number of different ways corresponding to 4 raised to about the 3-billionth power, a number which is far larger than the estimated number of elementary particles in the universe and which corresponds for our purposes to infinity. Furthermore, there seems to be no restriction on the order in which the bases can be arranged in the DNA molecule. No sequence is known to be impossible, but the elimination of all but a minute fraction of the possible sequences is carried out inexorably by the natural selection imposed upon the biological consequences of the linear order of the bases in DNA. These consequences are expressed by translation into protein molecules through the genetic code.

This thesis gives rise to several concepts. One is that the survival of organisms is possible only if their DNA molecules contain base sequences that can be translated into useful assemblies of proteins. Another is that the evolution of all living species is due to a series of changes in the order and length of the base sequences in DNA. A third is the deterministic conclusion that the purpose of life is the perpetuation of a base sequence; this perpetuation is brought about by a surrounding milieu of protoplasm, the nature of which is dictated by the base sequence itself. In effect, we strive, compete, and reproduce to maintain the continuity of half of a long word written in a four-letter alphabet, a word we are unable to read. The other portion of the word is, of course, contributed by the opposite sex.

These seemingly oversimplified and undoubtedly dogmatic concepts are being appropriately challenged and examined by a host of experimental scientists. The mechanisms involved in protein synthesis require elaborate techniques and much biochemical skill for their study in vitro. The field leads into a new area of aggregations of macromolecules, the organelles of the cell, which carry on many of the chemical tasks of life.

To recapitulate the postulates which have been discussed: All the necessary information for the perpetuation of each species is present in DNA in sequences of the four bases A, C, G, and T. When cells divide and multiply, this sequence replicates itself enzymatically through the Watson-Crick complementary pairing mechanism. In the everyday life of cells, portions of the sequence are translated into proteins through the genetic code. The translation results in definite series of amino acids being joined into specific proteins. The joining process is effected by formation of peptide linkages between adjoining amino acids; the carboxyl group of an amino acid joins to the α-amino group of a second amino acid and this procedure is repeated by the carboxyl group of the second amino acid joining with the α-amino group of a third amino acid thus:

$$NH_2-\underset{R_1}{\underset{|}{\overset{H}{\overset{|}{C}}}}-CO-NH-\underset{R_2}{\underset{|}{\overset{H}{\overset{|}{C}}}}-CO-NH-\underset{R_3}{\underset{|}{\overset{H}{\overset{|}{C}}}}-CO-\ldots-NH-\underset{R_n}{\underset{|}{\overset{H}{\overset{|}{C}}}}-COOH$$

This stepwise procedure is continued until the protein chain is completed. The order and number of the amino acids in each protein is directly dependent upon the sequences of bases in a corresponding portion of DNA, a *gene*. Evolution proceeds through stepwise changes, of an essentially accidental nature, in the number and sequences of bases in DNA, including complete or partial duplications of genes. Natural selection enables a small proportion of the changed molecules to survive because they produce groups of proteins that form viable organisms. Geographic separation, in terms of this concept, could well lead to the divergence of two groups of the same species, isolated from each other, through successive changes to a point where two new species will appear.

The information that leads to the characterization of each living species is contained in molecules of DNA. These molecules are known to differ from one species to another essentially only with respect to their length and the sequence of bases that they contain. The explanation for the mechanism of evolution must therefore be sought in changes that take place in the length of the DNA molecules and in the sequences of their bases.

This concept agrees with several experimental observations. The effects of environment are not inherited unless they produce changes in DNA. Actually, this axiom was perceived by Weismann (1892) long before the chemical properties of DNA were known:

> It is an inevitable consequence of the theory of the germ-plasm, and of its present elaboration and extension so as to include the doctrine of determinants, that somatogenic variations are not transmissible, and that consequently every permanent variation proceeds from the germ, in which it must be represented by a modification of the primary constituents....
>
> All permanent—i.e. hereditary—variations of the body proceed from primary modifications of the primary constituents of the germ.... Neither

injuries, functional hypertrophy and atrophy, structural variations due to the effect of temperature or nutrition, nor any other influence of environment on the body, can be communicated to the germ-cells, and so become transmissible.

We must exclude certain agents that produce changes in the molecules of DNA from Weismann's allusion to environmental effects. These agents include chemical mutagens, high temperatures, and ionizing radiation, which cause mutations to an extent that concords with their known effects on DNA.

There are also chemical agents that induce polyploidy, an actual increase in the amount of DNA per cell. Polyploidy leads to improvements in the vigor of plants, but in animals its occurrence produces disturbances in the chromosomal mechanism that determines sex, so that sexual reproduction is no longer possible.

In contrast, nonmutagenic changes, such as mutilations, are not inherited. Weismann furnished an experimental demonstration of this by amputating the tails of white mice through five generations and observing the offspring. The mutilated parents produced, of course, offspring with normal tails, an observation which is said to have led a contemporary biologist to quote the reflective statement by Hamlet:

> There's a divinity that shapes our ends
> Rough-hew them how we will.

The enormous numbers of possible changes in DNA provide a basis for natural selection, which was regarded by Darwin as the agency in evolution that led to the preservation of "favorable" variations and the rejection of "injurious" ones. Changes in DNA may lead to a failure of the cell to survive the next mitotic event or to modifications of an organism that are unable to survive, reproduce, or effectively compete. In a few rare instances, the changes may result in Darwin's favorable variations that are preserved through future generations, perhaps until they are displaced by the arrival of new species that can compete more successfully.

Another supportive observation is the fact that some of the same functional proteins are present in more than one living species. These proteins from several species have been completely analyzed, and the exact order in which their amino acids are linked together is known. When these sequences are compared, differences are seen to exist from one species to another, and these differences often roughly correspond to the relationship of the various species in the evolutionary family tree.

Still another example of evolutionary differentiation is found in the DNA molecules themselves. The percentage of the various bases in DNA is a rough guide to the taxonomy of bacteria (Marmur, Falkow, and Mandel, 1963) because in these organisms there is apparently an evolutionary tendency to move in the direction of either more or less $G + C$ in the total DNA. A more refined procedure is to study the formation of molecular hybrids by hydrogen bonding between the separated strands of DNA of two different species. Such hybrids indicate the existence of identical or closely similar base sequences in parts of the DNA molecules that are being compared. The hybrids are formed either in solution, in which case they are detected by centrifugal procedures, or on agar columns, in which case the hybrid molecules stay on the column while the unbonded single-strand fragments are washed off by an eluent (Hoyer, McCarthy, and Bolton, 1964). Such experiments, which will be discussed in a later chapter, are a means of measuring the relationships between species on the basis of similarity between DNA molecules. The significance of the results is not that they prove that a rat resembles a mouse, a conclusion which can be reached without recourse to the procedures of the biochemical laboratory, but rather that the findings reinforce the concept that evolution has proceeded by means of stepwise changes starting from archetypal DNA molecules and that the molecules have retained a great deal of similarity despite these changes. The great importance of the experiments is that they provide a direct demonstration of similarities between the base sequences of DNA molecules, a result which without the experiments is based solely on inferences derived principally from examining the amino acid sequences of proteins.

Evolutionists have not been slow to point out that the survival of the whole organism, based upon its interaction with the environment, is the predominant factor in natural selection, which cannot as a rule act on particular molecules in an analytically separable way (Simpson, 1964). It is not necessary, however, to propose, as Simpson (1964) has done, that there must be some sort of feedback from the organism-environment interaction into DNA, unless what is meant is that this interaction, through natural selection, eliminates many DNA molecules from their further participation in the self-replicative process.

The examples used in this book to illustrate the processes of evolution at the molecular level are, for the most part, proteins. Anfinsen (1959) drew on such examples in his book, *The Molecular Basis of Evolution*. If we consider as an example the two cytochrome c molecules found respectively in dogs and horses, it will be noted that these differ in about 10 of the amino acids in a chain of 104. The question arises, are the two molecules splendidly tailored to the different requirements specified by "dogfulness" and "horsefulness," have they evolved to conform to these two different requirements, or have the two cytochromes passively been carried along as dogs and horses evolved separately from a common ancestor? In the latter case, it is undoubtedly quite probable that separation of the two species would be followed by changes in the genes that in time would result in differences in the two cytochrome c molecules. We do not know whether a horse could get along just as well with dog cytochrome c as with its own. To cite another example, often quoted, the α chains of human and gorilla hemoglobins differ only with respect to seemingly inconsequential substitutions of glutamic acid by aspartic acid at site 23. It taxes the imagination severely to infer that this minuscule difference serves to distinguish the oxygen-transporting properties of the two hemoglobins in a manner that conforms with differences in the respective needs of the two species, so that the single amino acid difference has become fixed by natural selection.

It is also true, however, that human hemoglobins A and S are identical except for a glutamic acid/valine interchange at residue $\beta 6$; in this case a profound difference, repeatedly cited as being of evolutionary significance, is produced by the interchange.

The above examples indicate that the properties of each protein must be studied separately before the story of its evolution can be understood; each protein obviously represents only a tiny fraction of the related properties of a living organism. Proteins are the most variable and versatile of any class of biological molecules. The changes produced in proteins by mutations will in some cases destroy their essential functions, but in other cases the change allows the protein molecule to continue to serve its purpose. In contrast, DNA molecules are comparatively rigid, static, and inert and are functionally indistinguishable until their encoded messages have been phenotypically translated.

The difficulties of drawing conclusions about the *history* of the evolution of different species of living organisms from the examination of a few protein molecules are so obvious that they need not be enumerated. In contrast, the *processes* of evolution at the molecular level are greatly illuminated by studies of DNA, the synthesis and functions of ribonucleic acid (RNA), the amino acid code, protein synthesis, control mechanisms, and especially the structure and functions of proteins. Beyond these studies lie vast areas of biochemical complexity. The complexities appear to be due in large part to the numbers of intricate combinations of known components rather than to the presence of compounds whose nature is unknown.

[2]

The Genetic Code

> We have learned to bottle our parents twain in the yelk of an addled egg.
> Rudyard Kipling, "The Conundrum of the Workshops"

The facts of biology have thrust the coding concept upon science and have brought about its acceptance while its nature was still being explored; it would seem that if there had been no code, it would have been necessary to invent one.

The coding theory came into being in the wake of the "one gene–one enzyme" concept (Beadle and Tatum, 1941), which posed the question of the biochemical bridge between the information in the genes, which consist of DNA, and the structure of enzymes, which are made of amino acids joined in peptide linkages. The relationship had to take into account the fact that there are usually four variable units in DNA and that 20 different amino acids take part in the synthesis of proteins.

More than 20 amino acids are found in biological materials, but only 20 take part in protein synthesis (Yčas, 1958). These are listed in Table 2-1, together with their abbreviations, which are conveniently used in the depiction of polypeptide chains. When any one of the 20 is tagged isotopically, it can be shown that it is rapidly incorporated into proteins when introduced into a tissue preparation that is synthesizing protein. This contrasts with the behavior of amino acids that are not listed in Table 2-1; for example, it has been noted that C^{14}-hydroxylysine is not incorporated into collagen and that proline is a better precursor of the hydroxyproline residues of collagen than is hydroxyproline itself. Phosphoserine and diiodotyrosine are formed in proteins

Table 2-1. The 20 Amino Acids Concerned in Protein Synthesis and Their Abbreviations

#	Amino Acid	Structure	#	Amino Acid	Structure
1.	Glycine (gly)	$CH_2 \langle {}^{NH_2}_{COOH}$	12.	Tryptophan (try)	$\text{indole-}CH_2-CH\langle {}^{NH_2}_{COOH}$
2.	Alanine (ala)	$CH_3-CH\langle {}^{NH_2}_{COOH}$	13.	Proline (pro)	(pyrrolidine-COOH)
3.	Serine (ser)	$CH_2OH-CH\langle {}^{NH_2}_{COOH}$	14.	Aspartic acid (asp)	$HOOC-CH_2-CH\langle {}^{NH_2}_{COOH}$
4.	Threonine (thr)	$CH_3-CHOH-CH\langle {}^{NH_2}_{COOH}$	15.	Asparagine (asN)	$H_2NOC-CH_2-CH\langle {}^{NH_2}_{COOH}$
5.	Valine (val)	${}^{CH_3}_{CH_3}\rangle CH-CH\langle {}^{NH_2}_{COOH}$	16.	Glutamic acid (glu)	$HOOC-CH_2-CH_2-CH\langle {}^{NH_2}_{COOH}$
6.	Leucine (leu)	${}^{CH_3}_{CH_3}\rangle CH-CH_2-CH\langle {}^{NH_2}_{COOH}$	17.	Glutamine (glN)	$H_2NOC-CH_2-CH_2-CH\langle {}^{NH_2}_{COOH}$
7.	Isoleucine (ilu)	$CH_3-CH_2-CH-CH\langle {}^{NH_2}_{COOH} \atop \quad \quad \vert \atop \quad \quad CH_3$	18.	Histidine (his)	$\text{(imidazole)}-CH_2-CH\langle {}^{NH_2}_{COOH}$
8.	Cysteine (cys)	$HS-CH_2-CH\langle {}^{NH_2}_{COOH}$	19.	Arginine (arg)	${}^{H_2N}_{HN}\rangle\!\!=\!\!C-NH-(CH_2)_3-CH\langle {}^{NH_2}_{COOH}$
9.	Methionine (met)	$CH_3-S-CH_2-CH_2-CH\langle {}^{NH_2}_{COOH}$	20.	Lysine (lys)	$H_2N-(CH_2)_4-CH\langle {}^{NH_2}_{COOH}$
10.	Phenylalanine (phe)	$\text{C}_6\text{H}_5-CH_2-CH\langle {}^{NH_2}_{COOH}$			
11.	Tyrosine (tyr)	$HO-\text{C}_6\text{H}_4-CH_2-CH\langle {}^{NH_2}_{COOH}$			

Note: Asn, gln, ile, and trp are now used by the American Society of Biological Chemists rather than asN, gln, ilu, and try.

by phosphorylation and iodination of protein-bound serine and tyrosine, rather than being incorporated directly.

Arithmetical considerations require that a permutation of at least three of the nucleic acid units is needed to specify an amino acid in the translation process; a two-letter code can spell out only 16 words in an alphabet containing four different letters. This conclusion was pointed out by Dounce (1952), who proposed that the amino acids were specified by the "immediate surroundings" of the nucleic acid bases, so that there were 10 possible such surroundings for adenine (AAA, GAA, CAG, etc.) and 10 more for each of the other three bases, to give a total of 40 such triplets. If direction were of importance, the possibilities would be increased, and in any case "a sufficient number of nucleic acids could theoretically exist to account for the large number of proteins in nature, assuming that protein specificity is dependent upon amino acid arrangement in protein chains" (Dounce, 1952).

At that time there was no knowledge of the arrangement of bases in the molecule of DNA. In 1953, the crystallographic and model-building studies of Watson and Crick (1953a) and of Wilkins, Stokes, and Wilson (1953) showed that the genetic message in DNA could be written in the form of a linear sequence resulting from hydrogen bonding between the base pairs AT, TA, GC, and CG. The idea of the code was clearly stated by Watson and Crick (1953b):

> The phosphate-sugar backbone of our model is completely regular, but any sequence of the pairs of bases can fit into the structure. It follows that in a long molecule many different permutations are possible, and it therefore seems likely that the precise sequence of the bases is the code which carries the genetical information.

Table 2-2 shows the familiar abbreviations used in depicting DNA and RNA molecules.

The necessary corollary to this idea is that the code is translated into amino acid sequences in proteins and that these sequences are responsible, directly or indirectly, for all the phenotypic characteristics of living organisms. This second concept gained strength two years before the publications by Watson and Crick (1953a, b). Sanger (1952) determined experimentally the complete amino acid sequence of insulin and found that the same amino acids always occupied the same loci in the insulin from a given species. He and his collaborators also found that the insulins from two different mammals were very similar, differing often in only one or two amino acids at definite loci. Each insulin molecule contains two polypeptide chains linked by disulfide bridges.

Table 2-2. Abbreviations Used Commonly in Depicting Molecules of Deoxyribonucleic and Ribonucleic Acids (DNA and RNA)

Name	Abbreviation
Adenine, cytosine, guanine, and thymine deoxyriboside triphosphates	dATP, dCTP, dGTP, TTP (or dAppp, etc.)
Adenine, cytosine, guanine, and uracil triphosphates	ATP, CTP, GTP, UTP (or Appp, etc.)
Pentaribonucleotide fragment of RNA chain, beginning with 5′-phosphate and ending with 3′-OH	pApCpGpUpU or A C G U U (structure with P groups) —OH or A C G U U
Complementary Watson-Crick pairing in DNA with two strands of opposite polarity	A = T C ≡ G ↑ ↓ G ≡ C T = A T = A

The shorter of the two is termed the *A chain*, and the amino acids are numbered from left to right (Fig. 2-1). The amino acids in positions 8, 9, and 10 in the A chain and at position 30 in the B chain vary in different mammals. For example, threonine at locus A8, isoleucine at locus A10, and threonine at locus B30 in human insulin are replaced by alanine at locus A8, valine at locus A10, and alanine at locus B30 in beef insulin. It may be that, as much as any other single discovery,

a

b

Fig. 2-1. Amino acid sequences in bovine insulin: (a) as commonly depicted; (b) with deletions as postulated by Eck (1964). The numbers refer to the base changes corresponding to the differences in the coding triplets for the pairs of amino acids at corresponding loci in each chain.

these findings gave a clue to the biochemical basis of speciation because amino acid changes can directly reflect mutations in the bases of DNA.

It is interesting to reflect on the history of research on insulin, spurred, of course, by the need for a therapeutic answer to the problem of diabetes. This search led to the experiments by Banting and Best (1922), in which extracts of pancreatic tissue were found to alleviate the disease because of the presence of insulin in the extracts. The challenge then became to discover the chemical structure of insulin in the hope of producing it synthetically. Perhaps it was this train of thought and experimentation that led Sanger to determine its amino acid sequence. As a result, it was realized that proteins, which, prior to Sanger's studies, were often thought to be random arrangements of a characteristic group and number of amino acids, were actually linear sequences in which each amino acid had a definite place. A good analogy is the letters in the words on a page of a book. These must be in a fixed order and sequence for the words to "make sense," and each time the book is printed, the order is maintained, even when the type is reset.

End-Group Analysis

Each polypeptide chain may typically have an α-amino group at one end and a free α-carboxyl group at the other end, for example,

$$\begin{array}{cccc}
\text{H}_2\text{N}-\text{CH}_2-\text{CO}-\text{NH}-\text{CH}-\text{CO}-\text{NH}-\text{CH}-\text{CO}-\text{NH}-\text{CH}-\text{COOH} \\
 & | & | & | \\
 & \text{CH}_3 & \text{CH}_2 & (\text{CH}_2)_2 \\
 & & | & | \\
 & & \text{CONH}_2 & \text{S} \\
 & & & | \\
 & & & \text{CH}_3 \\
\text{gly} & \text{ala} & \text{asN} & \text{met}
\end{array}$$

Reagents that combine firmly with the end groups will serve two purposes: first, they will be "handles" for the identification of the terminal amino acids if the chains are broken down into individual amino acids by hydrolysis of the peptide linkages; second, the number of moles of such a reagent that reacts with a protein sample will indicate the number of moles of the protein in the sample and hence its molecular weight.

1-fluoro-2,4-dinitrobenzene (FDNB) combines with amino groups to yield dinitrophenylamine (DNP) compounds, which are yellow, as follows:

$$O_2N\text{-}C_6H_3(NO_2)\text{-}F + H_2N\text{-}CH_2\text{-}COOH \xrightarrow{NaOH}$$

$$O_2N\text{-}C_6H_3(NO_2)\text{-}HNCH_2\text{-}COOH + NaF$$

A protein may be allowed to react with FDNB in alkaline solution and then hydrolyzed with strong acid. The DNP amino acids are then separated by extraction and chromatography. These are α-amino acids with a free NH_2 group at the end of peptide chains, such as glycine in the example above. Lysine has a second amino group, the ε group, which is free whether or not the α amino group of lysine is held in peptide linkage. The ε amino group on the side chain of lysine forms a DNP compound that is separable from the α DNP compounds by solvents.

Phenylisothiocyanate is another reagent that reacts with amino-terminal groups. It does so in the following manner:

$$C_6H_5\text{-}NCS + H_2NCH_2CONH\text{-}R \quad \text{(peptide chain)}$$

$$C_6H_5\text{-}NHCSNHCH_2CONH\text{-}R$$

$$\begin{array}{c} C_6H_5\text{-}N\text{---}CS \\ || \\ OCNH \\ \diagdown\diagup \\ CH_2 \end{array} + NH_2\text{-}R \quad \text{(remainder of peptide chain)}$$

Phenylthiohydantoylglycine

This procedure may then be repeated with the next free-amino-terminal group. This is the Edman stepwise degradation.

In some proteins the terminal amino group exists as –NH-acetyl rather than $-NH_2$ owing to enzymatic acetylation following protein synthesis.

The other end of the polypeptide chain, ending in a carboxyl group, may be identified by hydrazinolysis, which liberates the carboxyl-terminal amino acid in the free state and converts the others to hydrazides. End groups may also be identified by the exopeptidases, amino-peptidases, and carboxypeptidases which liberate single amino acids one at a time from the respective ends.

Formation of Peptides by Enzymatic Hydrolysis of Proteins

This procedure is used to break the large protein molecule into characteristic fragments small enough to be identified with respect to their content and sequence of amino acids. The findings are then reassembled, like the pieces of a jigsaw puzzle, into the complete picture of the original protein.

Trypsin hydrolyzes the bonds between the carboxyl groups of lysine and arginine and the next amino acid in the peptide sequence. Chymotrypsin attacks the bonds that join the carboxyl groups of tryptophan, phenylalanine, and tyrosine—and, to a lesser extent, leucine, methionine, histidine, and asparagine—to the amino group of the next adjacent amino acid. The peptides that are produced by these and other proteolytic enzymes are separated from each other by column chromatography or by two-dimensional paper chromatography (*fingerprinting*).

Let us suppose that a peptide containing nine amino acids yields the following fragments, in which each letter represents an amino acid: PK, FK, DS, KS, FD, SE, SF, PKS, KSF, FKF, PKSF, FDSE. The sequence must be PKSF–FDSE or FDSE–PKSF. However, the sequence FKF is present. Therefore, the 9-peptide must be PKSFKFDSE. This example shows how information from the breakdown of proteins is put together.

Sanger and his collaborators found that insulin contained 51 amino acid residues and two terminal NH_2 groups. It therefore contained two peptide chains. These were shown to be held together by cross-linkage between the SH groups of cysteine residues, which can form –S–S– bridges. When these bridges were broken down by oxidation with performic acid, two peptides were obtained, one with NH_2-terminal glycine and the other with NH_2-terminal phenylalanine. These were degraded by enzymatic and chemical procedures, and the sequences were reassembled by the methods described above. The complete structure of the molecule obtained from beef pancreas was found to be as shown in Fig. 2-1.

An examination of the structure of beef insulin shown in Fig. 2-1*a* shows another evolutionary relationship. The two peptide chains of the insulin molecule are sequentially related in a manner which indicates that they arose by partial gene duplication and subsequent evolution that involved changes in the bases in the genes as suggested by Gamow, Rich, and Yčas (1956). Eck (1964) has made the further suggestion that, in addition to substitutions, deletions have occurred since the duplication took place. This relationship between the two peptide chains is shown in Fig. 2-1*b*. The deletions are located by juxtaposition of identical or "related" amino acids in the two chains. It is thought that pieces of DNA, each piece containing a multiple of three base pairs, have been lost from the gene during evolution. In Fig. 2-1*a* the row of figures beneath the lower peptide chain indicates the minimum number of base changes, total 32, needed following duplication to bring about the differences in amino acids between the polypeptides coded by the two genes. In Fig. 2-1*b* deletions are assumed to have taken place as indicated by –. In this case, the minimum number of base changes involved in the changes is only 13, and, perhaps even more important, nine sites are unchanged as compared with only two in Fig. 2-2*b*. These considerations support the proposal by Eck. We shall discuss this point in Chapter 6.

This discussion has shown us how the gene places its imprint on a protein, which is a chemical molecule that may be obtained in large quantities in pure form for the determination of its structure. We now return to a further consideration of the relation between DNA and amino acid sequences in proteins. Various theoretical proposals have been made to explain the process of translating the base sequence of the gene in the DNA molecule into the amino acid sequence of the protein which was coded by the gene. The first such proposal was by Gamow (1954*b*), who suggested that proteins were formed on the surface of the DNA double helix. The amino acids were thought to find their way into "holes" between the bases. The surrounding of each hole was specified by the four bases at its corners:

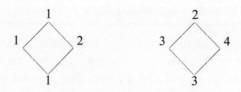

Two of the opposite corners were always occupied by a complementary pair of bases.

This scheme imposed certain restrictions upon amino acid sequences, and examination of the primary structure of a number of proteins later showed that any amino acid may be next to any other, so that these restrictions do not exist (Brenner, 1957). The difficulties of accommodating amino acids in the alleged "holes" in the DNA molecule and the doubt that protein synthesis takes place in the nucleus of the cell were further problems that confronted the scheme.

Other proposals were made by Gamow (1954a), Gamow and Yčas (1955), Crick, Griffith, and Orgel (1957), and Woese (1961). Crick et al. (1957) derived sets of 20 triplets, each consisting of a three-letter permutation in a four-letter alphabet, in order to satisfy a criterion that each of the 20 amino acids involved in protein synthesis should correspond to a three-base triplet and that one could place any two triplets next to each other without producing overlapping triplets which belong to the set. Their scheme excluded triplets consisting of the same letter taken three times as codes for any amino acid. Such triplets, such as AAA, have known coding assignments.

Viewed from the advantageous standpoint of retrospect, the coding proposals made prior to 1961 did not have access to one or more of the considerations used in current thinking. Some of these are as follows:

1. The sequence of the four possible base pairs in the DNA molecule (AT, TA, GC, CG) is unrestricted, for all practical purposes, with respect to frequency of "nearest neighbors." Although certain trends in such frequencies have been recorded (Josse, Kaiser, and Kornberg, 1961), they have not been shown to be correlated with restrictions upon

the sequencing of amino acids in proteins; indeed, no such restrictions are known to exist.

The reservation must be made that apparently most proteins so far examined *start* with an amino acid that is coded by a triplet starting with a purine, presumably because purine nucleoside triphosphates are used principally or exclusively to start RNA chains in the DNA-dependent RNA polymerase reaction (Maitra and Hurwitz, 1965; Krakow, 1965). It seems also that some proteins may be initiated with a formyl-methionine molecule which is later removed (Watson et al., 1966; Zinder et al., 1966).

There is, however a definite although unknown restriction in the overall amino acid sequence in each naturally occurring protein molecule. This restriction is imposed by the fact that each protein must have a definite secondary and tertiary structure to make it fit for its biological tasks. The secondary and tertiary structures arise directly from the amino acid sequence. Certain variations are permissible in the amino acid sequences that are compatible with the required structure. These variations are produced during the evolution of various living species.

Spetner (1964) has examined the problem of the role of trial and error in evolution. Natural selection is considered as a mechanism which transfers information from the environment to the gene in the sense that random mutation produces an assortment of genotypes of which one or more is favored by the environment. He has calculated the probability P that in N trials (generations) or less a given polynucleotide of length n will mutate into any one of m specified sequences of length u. This probability not unexpectedly is a strong function of p, the mutation probability per nucleotide, assuming that the only mutations are changes in single nucleotides.

2. It is assumed that 60 to 62 of the 64 possible triplet sequences of four bases can specify the code for a single amino acid and that there is no appreciable ambiguity of the specificity with which a triplet selects an amino acid, although ambiguities are sometimes observed under test-tube conditions (Davies, Gorini, and Davis, 1965), especially in the presence of streptomycin.

3. The fact that some amino acids are more common (e.g., alanine,

leucine, glycine) than others (such as cysteine and tryptophan) is not necessarily because the more common amino acids have a larger number of alternate or synonymous coding triplets than the rarer ones. Evolutionary selection of proteins and hence of amino acids could play a part in deciding the average frequency of a coding triplet in the genetic DNA.

4. The selection of the amino acids at the time and place of polypeptide synthesis is carried out by a single strand of "messenger" RNA (Jacob and Monod, 1961). This is produced by an enzymatic reaction in which only one of the two strands of DNA is used as a template in the formation of a complementary RNA strand. It is easy to perceive that a second strand, produced with reverse complementarity from the other strand of DNA as a template, would almost certainly not produce a biologically acceptable protein, in view of the criteria for proteins laid down in (1) above. Even if it were acceptable at some point in evolution, its acceptability would disappear as a result of mutational changes in DNA that could be tolerated only in the other protein coded by the forward complementary sequence. These tenets are strengthened by the finding that the enzyme RNA polymerase produces RNA complementary to only one of the two strands of the DNA template (Hayashi, Hayashi, and Spiegelman, 1963; Green, 1965).

5. There is evidence for the participation of RNA in the synthesis of polypeptide chains. The mechanisms involved will be described below. The chains are formed by sequential, linear, and unidirectional addition of amino acids in peptide linkage. The first amino acid to be incorporated is the one that has a free amino group in the finished chain, and the final amino acid in the sequence has, of course, a free carboxyl group. This was ingeniously demonstrated by Dintzis, who examined the relative amounts of radioactivity in the tryptic peptides of hemoglobin following the addition of a "pulse" label of a tritium-tagged amino acid to a preparation of rabbit reticulocytes that was actively synthesizing hemoglobin (Dintzis, 1961; Naughton and Dintzis, 1962). The preparation contained incomplete chains of hemoglobin when the tritium pulse was added. Time was allowed for the incomplete chains to

be finished off. Analysis of the chains showed that more of the radioactivity was at the carboxyl end; therefore this end was made last.

6. The amino acid code is shown to be "nonoverlapping" because many mutations have been shown to consist of a single amino acid change in a protein, and in each case the two *nearest-neighbor* amino acids are not changed. In an overlapping code, a single-base change would often change either two or three amino acids by changing the bases in two or three triplets. Furthermore, some of the single-amino-acid mutations have been produced by treating viral RNA with chemical reagents such as nitrous acid which are known to produce changes in purine and pyrimidine bases, such as adenine to hypoxanthine and cytosine to uracil. The adjoining amino acids are unchanged. This is additional evidence that a single-base change causes a single-amino-acid change.

7. "Artificial" messenger RNA, consisting of enzymatically synthesized polyribonucleotide molecules, will serve as a template for the production of polypeptide chains in cell-free preparations. The composition of the polypeptides thus formed is directly related to the base composition of the polyribonucleotides.

8. Trinucleotide molecules each containing a linear sequence of three bases, pXpYpZ or XpYpZ, were shown to bring about the binding of specific sRNA molecules to ribosomes (Nirenberg and Leder, 1964).

There is much evidence that protein synthesis occurs principally in the cytoplasm of cells rather than in the nuclei and that it takes place in association with the small particles of ribonucleoprotein that are termed *ribosomes*. Ribosomal RNA is fairly constant in composition, and its base ratios do not correspond with those of DNA, which vary greatly among species. This finding was a stumbling block in rationalizing the composition of proteins with the base sequence in DNA. The difficulty was resolved by the discovery of an unstable form of RNA in the cytoplasm which had a base composition similar to that of the DNA. This led to the widely accepted "messenger" hypothesis (Jacob and Monod, 1961), which proposes that the base sequence in DNA is transcribed into a strand of messenger RNA, the molecule along which the amino acids are assembled in the correct sequence during the synthesis of polypeptide strands of which proteins are composed.

Another problem existed: amino acids in the free state are unable to recognize the sequence of bases in messenger RNA, and, as we have seen, each amino acid would have to recognize one or more triplet sequences. A solution to this problem emerged from studies of the procedure in which amino acids are "activated" for participation in protein synthesis. These investigations (Hoagland, 1955; Hoagland, Keller, and Zamecnik, 1956; Berg, 1956) substantiated Crick's "adaptor hypothesis" (1958).

The activation of amino acids takes place in two steps. The first is a reaction between an amino acid and adenosine triphosphate (ATP), with an enzyme that is specific for the amino acid, to form an amino acyl–AMP–enzyme compound with the liberation of pyrophosphate. This compound now recognizes another specific molecule, a molecule of transfer RNA (sRNA), which contains about 77 bases, some of them methylated. The 77 bases include also a few pseudouridine (5-ribosyluracil) groups. The sRNA is the adaptor molecule which will locate the correct site of three consecutive bases on the messenger RNA corresponding to the code for the amino acid. The activating enzyme and AMP are now released from the *charged* sRNA, which carries the amino acid attached through the 2'- or 3'-OH group of the ribose portion of its "right-hand" terminal nucleotide. This is always adenylic acid, and the next two nucleotides in the chain are both cytidylic acid residues (Fig. 2-2). This –CCA group is added to each sRNA molecule by a special reaction.

The union of the activated amino acid with sRNA is the point of transfer of information between the nucleic acid alphabet of four letters (the bases) and the protein alphabet of 20 letters (the amino acids). The recognition must be between the amino acid molecule and a specific base sequence termed the *recognition site* on the sRNA molecule and must be mediated by an amino acid sequence in the specific activating enzyme. It is quite probable that there are two essential sites on the enzyme, one of which fits the amino acid and the other the sRNA recognition site. In any event, this procedure, the charging of sRNA, is one of the most intricate and vital procedures in the chemistry of life.

What presumably follows is somewhat less complex by comparison. The sRNA molecule contains a group of three bases, the coding site, or *anti-codon*, which, aided by "transfer enzymes," seeks out its reverse

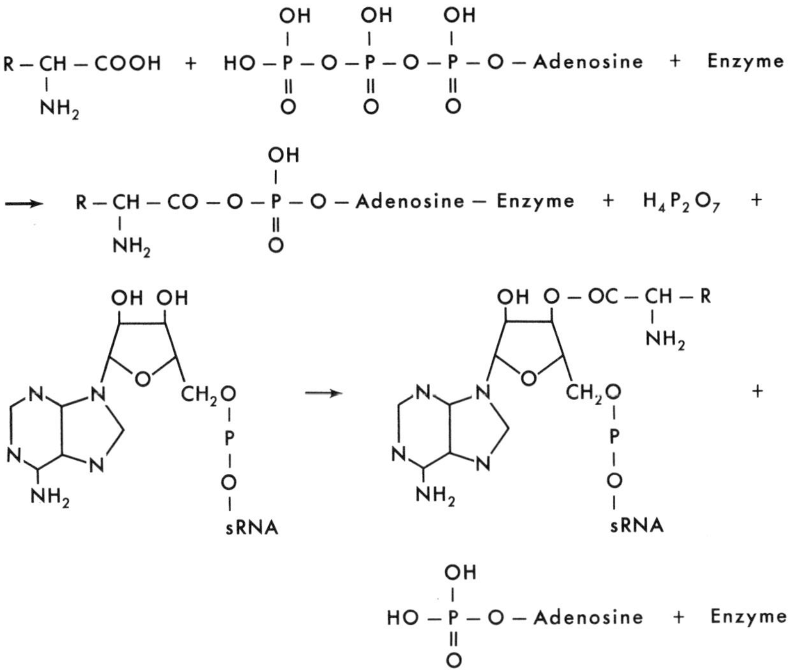

$$R-CH(NH_2)-COOH \;+\; HO-\underset{O}{\overset{OH}{P}}-O-\underset{O}{\overset{OH}{P}}-O-\underset{O}{\overset{OH}{P}}-O-\text{Adenosine} \;+\; \text{Enzyme}$$

$$\longrightarrow \; R-CH(NH_2)-CO-O-\underset{O}{\overset{OH}{P}}-O-\text{Adenosine}-\text{Enzyme} \;+\; H_4P_2O_7 \;+\;$$

Fig. 2-2. Amino acid activation.

complement on a strand of messenger RNA that is bound to the 30S particle of a ribosome. The procedure has often been shown in diagrammatic form, as in Fig. 2-3b. The messenger travels over a series of ribosomes; the combination is called a *polyribosome*, or *polysome* (Fig. 2-3a). Each ribosomal unit in an actively functioning polysome

26 The Genetic Code

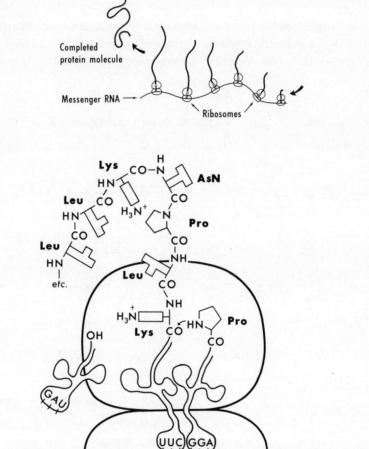

Fig. 2-3. (*a*) Synthesis of protein on a polysome. (*b*) Interaction of messenger RNA with transfer RNA on a ribosome, showing the progress of polypeptide synthesis. The discharged transfer RNA molecule on the left has just been released from the ribosome.

has a growing polypeptide chain attached to it by means of an sRNA molecule. The shortest chain is on the ribosome which has most recently become attached to the messenger. The longest polypeptide chain is on the ribosome at the other end of the polysome. The ribosome is released when it reaches the 3'-OH terminal end of the messenger strand, and simultaneously the completed polypeptide molecule is set free. The coding site is distinct from the recognition site. This distinction actually makes possible the existence of one sRNA for each coding triplet in messenger RNA, and it is customary to regard the coding triplet in sRNA as the determining factor in distinguishing the sRNAs from each other. There could be at least four sRNA molecules specific for alanine, containing the coding sites, AGC, GGC, CGC, and UGC, and each containing the same recognition site. The presence of IGC, however, as the presumed coding triplet in yeast alanyl sRNA has given rise to speculation that IGC may pair with the alanine codes GCA, GCC, and GCU in mRNA (Crick, 1965), especially in view of the finding that the trinucleotides GCA, GCC, and GCU will bind the same yeast alanyl sRNA to ribosomes (Brimacombe et al., 1965). Dütting and coworkers (1965) have recently reported the presence of IGA in yeast seryl sRNA, which could be interpreted similarly in terms of the serine codes UCA, UCC, and UCU (Tables 2-12 and 2-16).

The evolution of the sRNA–activating enzyme–amino acid–ribosome–messenger RNA system and its maintenance from generation to generation through the ages by means of information stored in the base sequence of DNA are indeed impressive phenomena. The elegance and intricacy of the processes excite one's admiration. It is noteworthy that they have been discovered within the last decade as a result of the fact that they have existed continuously for several hundred million years. The processes have thus given rise, by evolution, to the human species and in consequence to their own revelation.

The evidence that the order of bases in DNA molecules is directly related to the sequence of amino acids in proteins is inferential. There is no known way of separating and identifying a piece of DNA in terms of its genetic function. This contrasts markedly with the opportunities

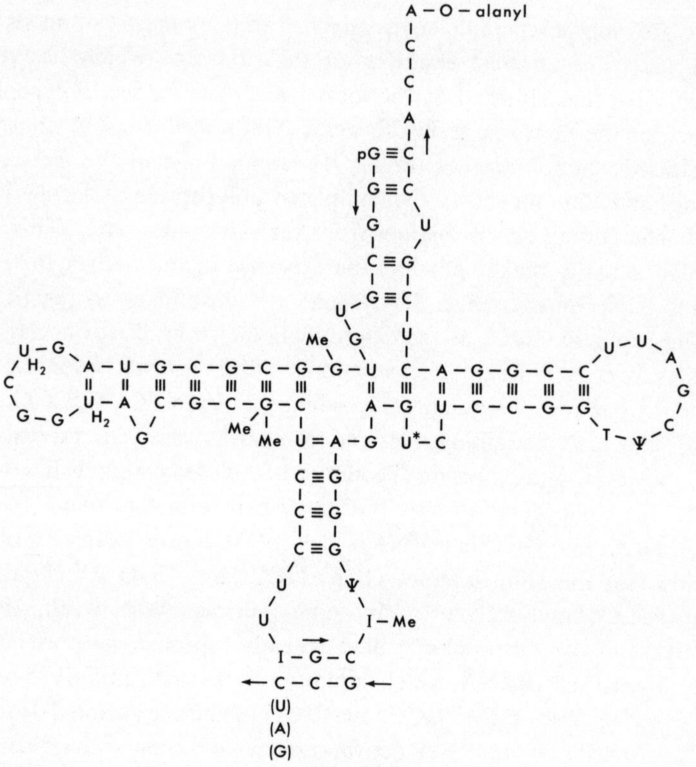

28   The Genetic Code

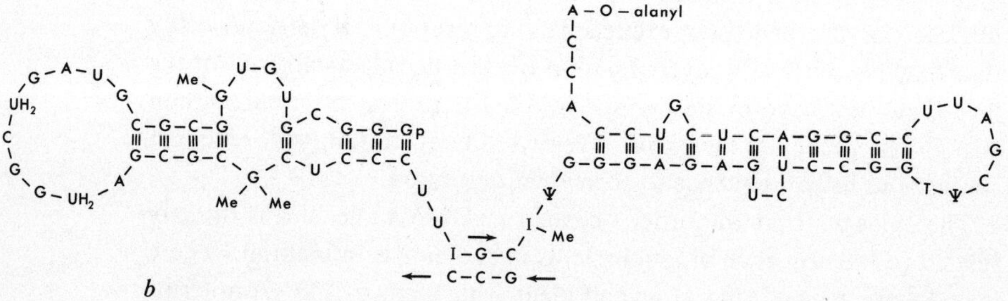

Fig. 2-4(a) and (b). Possible arrangements of the base sequence in yeast alanyl sRNA, showing possible complementation of its coding triplet with alanine codes in mRNA. Arrangement (a) is from Tinoco (1965).

available for characterizing proteins, which can be purified and crystallized, often in large quantities, as discrete entities.

In the remainder of this chapter we shall discuss some questions related to the base composition of DNA and RNA and some of the processes by which RNA brings about the expression of the genetic message into phenotypic characteristics.

The determination of the order of bases in DNA is extremely difficult, and very little progress has been made in this field; indeed, Chargaff (1955) characterized the task as so gigantic as to discourage the most sanguine of optimists. The molecule is of enormous length in relation to its diameter and easily becomes inadvertently broken by manipulation. As soon as it is broken, the sequence is fragmented, so that the beginning and end are no longer identifiable. Even in the case of RNA, which exists in much shorter molecules, there is very little information with respect to base sequences. The smallest molecules in the RNA family are the sRNA molecules. These have –CpCpA groups at the 3' end and pG– at the 5' end. The –CCA group is added after the chain has become detached from its site of synthesis. A few of the bases that are close to the –CCA have been determined by Berg, Lagerkvist, and Dieckmann (1962) for some of the sRNA molecules. A number of oligonucleotides from digests of sRNA have been identified by various investigators (Armstrong et al., 1964; Cantoni et al., 1963; Dütting et al., 1965; Holley et al., 1965; Ingram et al., 1964).

Holley et al. (1965) have described their findings for the complete sequence of bases in an alanyl-sRNA prepared from yeast. This is shown in Fig. 2-4, which has been drawn in two of the various forms illustrative of possible cross-linkage between complementary pairs of the 77 bases in the molecule. The sequence IGC, equivalent to GGC, in the loop at the bottom of the diagram, could, as noted earlier, conceivably furnish the coding sequence that seeks out and complements with GCC, GCU, and GCA (Crick, 1965) (read in reverse) in messenger RNA, the triplets for three alanine codes. The remainder of this remarkable molecule presumably include sites that are involved in binding to the ribosome and in recognition of the functional site of the alanine-activating enzyme. Little or nothing is known of the order of bases in the larger RNA molecules.

The average base composition of DNA molecules was studied by Chargaff and his collaborators (1955), who first showed that the molar percentages of adenine and thymine were identical and that those of guanine and cytosine were also identical (percent A = percent T and percent G = percent C). The purines may be removed from chains of DNA by mild acid hydrolysis, so that apurinic acids, containing runs of consecutive pyrimidines are obtained (Tamm, Modes, and Chargaff, 1952). By such procedures poly-T sequences up to five (subsequently seven) nucleotides in length were obtained (Spencer and Chargaff, 1962), and it has been pointed out that poly-T sequences are longer than poly-C sequences (Rudner, Shapiro, and Chargaff, 1962), which so far have been found only up to 3C in length. In terms of coding assignments (Table 2-12), the sequence of lys-lys-lys-ilu, which is found in cytochrome c, could reflect the presence of a 10-T sequence in DNA.

The percentage of the various possible base pairs ("nearest neighbors") in the DNA of different species was extensively explored by Kornberg and his coworkers by the following procedure: DNA of a specific origin was used as a template in the DNA polymerase reaction, and one of the four deoxynucleoside triphosphates in the reaction mixture for building the new chains was prepared with P^{32} in the phosphate esterified to the sugar in the 5′ linkage. The enzymatic synthesis produced chains of polynucleotides with some of their diester bonds containing radioactive phosphorus. The chains were then degraded enzymatically to mononucleotides by micrococcal deoxyribonuclease and spleen phosphodiesterase. This procedure breaks the linkages which unite phosphorus to the 5′ position of deoxyribose but leaves the linkages to the 3′ position intact, so that, in effect, the P^{32} is shifted from the original nucleoside to its nearest neighbor (Josse et al., 1961). In abbreviated form, this may be written as

$$X_p Y_p Z + \text{ppp}^*T \rightarrow X_p Y_p Z_p{}^*T + \text{pp}$$
$$X_p Y_p Z_p{}^*T \xrightarrow{\text{spleen diesterase}} X_p + Y_p + Z_p{}^* + T$$

The analytical results yield the frequency of the 16 possible nearest-neighbor arrangements in the DNA chains as follows: Radioactive thymidine triphosphate, abbreviated as ppp*T, is used, together with the cold 5'-triphosphates of the other deoxynucleosides in the DNA polymerase reaction, employing, for example, calf thymus DNA as a primer. The product is then degraded to deoxyribonucleotides with the phosphate in the 3'-position, namely, Ap, Cp, Gp, and Tp. These are assayed for radioactivity. Every radioactive molecule of Ap* must have been next to p*T, so that the proportion of Ap* in the hydrolysate represents the proportion of ApT groups in the DNA. Similarly, the percentages of CpT, GpT, and TpT are calculated. The procedure is then repeated with ppp*C to find the relative amounts of ApC, CpC, GpC, and TpC, and so on for ppp*A and ppp*G.

If the polarity of the two complementary chains of DNA runs in opposite directions, TpA in chain 1 read in one direction is also TpA in chain 2 read in the opposite direction, for TpA complements ApT. The same rule holds for ApT, CpG, and GpC. The other 12 pairs do not exhibit identity as a function of reverse complementarity, and the amount of ApA should equal TpT, CpA should equal TpG, etc. These relationships were confirmed for a large number of DNA preparations obtained from double-stranded templates, thus supporting the anti-parallel-strandedness concept of DNA. An example is shown in Table 2-3. Each source of DNA led to products which had a unique and nonrandom pattern of the 16 nearest-neighbor frequencies; however, the values for the DNA of three of the T-even bacteriophages, T2, T4, and T6, were virtually identical, as would have been expected from their close genetic similarity. This similarity was subsequently emphasized by the fact that single strands prepared from these DNA molecules would readily hybridize (Schildkraut et al., 1962) with their antiparallel counterparts from another T-even phage.

The percentages of each doublet in DNA, obtained from the nearest-neighbor experiment, may be used to calculate the relative probabilities of the 64 triplets (Gatlin, 1963). This procedure does not yield the relative probable amounts of triplets that actually take part in coding, since the figures thus obtained are the averages of both strands and the messenger RNA is the complement of only one strand; for example, the total value for GpApA in both strands represents the sum of the values of GpApA (glu) and TpTpC (phe) in either strand. It is possible to tabulate the relative probable amounts of each triplet in the *single-stranded* DNA of ϕx-174 bacteriophage by this method. These values

Table 2-3. Nearest-Neighbor Frequencies of Bases in DNA Enzymatically Synthesized Using Primers from Two Bacterial Sources

Nearest-neighbor sequence:	*M. lysodeikticus* DNA[a]	*B. subtilis* DNA[b]
ApA, TpT	0.019, 0.017	0.092, 0.095
CpA, TpG	0.052, 0.054	0.067, 0.068
GpA, TpC	0.065, 0.063	0.067, 0.065
CpT, ApG	0.050, 0.049	0.057, 0.058
GpT, ApC	0.056, 0.057	0.048, 0.048
GpG, CpC	0.112, 0.113	0.046, 0.046
TpA	0.011	0.052
ApT	0.022	0.080
CpG	0.139	0.050
GpC	0.121	0.061
Deviation ratios:		
$\sigma_{opp}/\sigma_{error}$	0.6	0.5
$\sigma_{sim}/\sigma_{error}$	6.7	7.4
$\sigma_{rand}/\sigma_{error}$	5.8	6.7

[a] Base composition: 14.7 percent A, 14.5 percent T, 35.4 percent G, 35.4 percent C.
[b] Base composition: 27.8 percent A, 28.0 percent T, 22.2 percent G, 22.0 percent C.
Source: Josse et al. (1961).

in terms of their complements should directly reflect the composition of the ϕx-174 messenger RNA.

Sueoka (1961c) compared the base analyses of the DNA of bacteria with the percentages of amino acids in their total protein in an attempt to find a relationship between the base content and the amino acid code. Bacteria are remarkable for exhibiting a wide range of base ratios, extending from 72 percent G + C for *Micrococcus lysodeikticus* and certain *Streptomyces* to 34 percent G + C for *Bacillus cereus*. The amino acid percentages in the total protein of various bacteria vary within much narrower limits; bacterial protein is characteristically high in the "primitive" amino acids (p. 68), especially glycine, alanine, valine, and leucine, and low in histidine, tyrosine, methionine, and tryptophan. Indeed, the amino acid percentages vary so little that it is possible to speculate that *M. lysodeikticus* has evolved in a manner

Table 2-4. Evolution of DNA of *M. lysodeikticus* and *B. cereus* from Archetypal Doublet Code to Present Base Ratios

Amino Acids	Doublet (Archetypal Code)	*M. lysodeikticus* C or G triplets			*B. cereus* U or A triplets	
		mol % Amino Acid	GC, %		mol % Amino Acid	GC, %
Primitive:						
ala	GC	16.4	16.4		10.0	6.7
arg	CG	5.5	5.5		4.3	2.8
asp + glu[a]	GA	10.8	7.2		11.0	3.7
cys	UG	...	...		0.3	0.1
gly	GG	11.1	11.1		9.6	6.4
his	CA	1.6	1.0		2.0	1.3
ilu	AU	3.9	1.3		6.8	...
leu	CU	8.4	5.6		8.8	2.9
lys	AA	4.4	1.5		6.1	...
phe	UU	2.4	0.8		3.8	...
pro	CC	4.7	4.7		3.6	2.4
ser	AG	4.0	2.6		4.3	1.4
thr	AC	5.8	3.9		5.6	1.9
val	GU	8.4	5.6		7.7	2.6
	Triplet				Triplet	
Recent:						
asN[a]	AAC	4.5	1.5	AAU	4.8	0
glN[a]	CAG	6.3	4.2	CAA	6.2	2.1
met	AUG	0.8	0.3	AUG	1.9	0.6
try	UGG	...	...	UGG	...	...
tyr	UAC	1.4	0.5	UAU	3.1	0
Total		100.4	73.7		99.6	34.9

[a] Extrapolated as 50 percent of Asx and Glx, respectively.

reminiscent of the effect of the "mutator gene" that brings about AT to CG (and TA to GC) changes in the base pairs of DNA (Yanofsky et al., 1966), these changes being transcribed in the third base of triplets for ala, gly, val, etc. (Tables 2-4 and 2-13).

Sueoka was able to show highly significant correlations between base composition and amino acid content of bacterial protein in the following cases:

A + T correlated with isoleucine, lysine, phenylalanine, and tyrosine.
G + C correlated with alanine, arginine, and glycine.

These correlations are in accordance with the coding assignments for these amino acids (Table 2-12), and these are still the only findings that show any direct relationship between the base composition of DNA and the amino acid content of proteins.

For a further discussion of the expression of the genetic message, we must turn to a brief examination of the functions of RNA. RNA is formed in cells by the complementary translation of the base sequence of one strand of DNA into a strand of RNA by the enzyme *RNA polymerase*.

Indications of such a process were noted by Volkin and Astrachan (1956) in experiments with T2-phage-infected *Escherichia coli*. They found that a short-lived form of RNA was produced in such cells with a base distribution similar to that of the phage DNA, so that it is sometimes termed *DNA-like RNA*. This phenomenon is currently regarded as a step by which the phage usurps the protein-synthesizing functions of the host by the manufacture of a new template. The template furnishes the information leading to the production of proteins that will carry out the procedures necessary to make new units of bacteriophage.

The "new," unstable form of RNA was shown to be associated with ribosomes and to have a molecular weight of about 1.5×10^5 (Nomura, Hall, and Spiegelman, 1961; Brenner, Jacob, and Meselson, 1961). It was investigated for evidence of homology between its base sequence and that of DNA from bacteriophage T2. Heat denaturation separates the two strands of DNA, and if the hot solution is cooled rapidly, the reassociation of the strands is minimized. The hydrogen bonds between the base pairs adenine-thymine and cytosine-guanine in the two complementary strands of DNA are broken by heating, so that the strands become separated and random coils are formed (Marmur and Doty, 1961). The denatured DNA may be examined to test whether either of the strands in it will exhibit complementary binding with another strand of closely similar base sequence (Marmur and Lane, 1960). The length of the strands and the heterogeneity of the base sequence are so great that this would seem to make it most unlikely that two complementary strands, once separated, could find their way back together during cooling, so that the double strands would be reconstituted with the base pairs once more hydrogen-bonded, A to T and G to C, as before. This actually occurs, however, and the fact that it does so is the genesis of procedures for preparing various types of complementary double strands. The extent to which such double strands are formed between pieces of single-stranded DNA from various species is roughly correlated with their taxonomic relationships (Chapter 7).

A strand of RNA will complement with its parental single strand of DNA, which has served as a template for the formation of the RNA by the action of RNA polymerase. This enzyme pairs the riboside triphosphates of A, C, G, and U with their complementary deoxyribotides in a DNA template. A hybrid RNA-DNA double strand is produced. The resulting hybrid is resistant under certain conditions to the action of pancreatic ribonuclease, which may be used to remove the loose ends of the RNA strands. The hybrid double strands have been used to demonstrate the existence of regions complementary to various molecules of RNA in various segments of the DNA genome. Hybrids were formed in cesium chloride gradient suspensions in the ultracentrifuge between P^{32}-labeled RNA prepared from *E. coli* shortly after infection with T2 phage and heat-denatured H^3-labeled DNA from T2 phage (Hall and Spiegelman, 1961).

A more convenient and versatile procedure for investigating the complementarity between DNA and RNA was devised by Bautz and Hall (1962), who prepared chemically linked cellulose columns with heat-denatured DNA from bacteriophage T4. These columns enable "T4-specific RNA" to be separated from *E. coli* RNA by binding with the T4 DNA. The RNA is subsequently removed by elution after the two strands of the hybrid are separated. (See Chapter 7.)

RNA Polymerase

Enzyme systems that catalyze the formation of polyribonucleotides from the riboside triphosphates of A, C, G, and U were described by various investigators during 1960 and 1961 (for references see Krakow and Ochoa, 1963). DNA functioned as a template to determine the sequence of bases. Double-stranded DNA was more effective than the single-stranded form (Nakamoto and Weiss, 1962). Polyribonucleotides could also function as primers (Burma et al., 1961); poly-U was the most active. Homopolynucleotides directed the incorporation of the complementary nucleotides so that poly-A was formed when poly-U was used as a primer.

The antibiotic actinomycin D is a very potent inhibitor of the DNA-directed polymerase reaction but has no effect on the reactions of RNA polymerase with poly-A, poly-C, or poly-U as templates. Reich and Goldberg (1963) studied the mechanism of this reaction. They conclude that actinomycin D combines with guanine deoxyriboside in double-stranded DNA and directly inhibits RNA polymerase by blocking those surfaces of the DNA template that are involved in the reaction with this enzyme. Actinomycin does not inhibit DNA polymerase at the low concentrations which suffice to block RNA polymerase, nor does it inhibit the replication of single-stranded RNA viruses. It is therefore used in investigations of gene-blocking, since it stops the production of RNA by the action of RNA polymerase, and in studies of the rate at which RNA is broken down. For example, it was found that the production of all forms of RNA was suppressed in *B. subtilis* by actinomycin (Acs, Reich, and Valanja, 1963); this would refute the proposal that sRNA, unlike messenger RNA, serves as a template for its own replication (Spencer et al., 1962). Actinomycin rapidly inhibits the formation of protein in bacterial cells, indicating that in this case messenger RNA is short-lived and is soon broken down by ribonucleases. Actinomycin does not stop the synthesis of hemoglobin in reticulocytes (Reich et al., 1962), the synthesis of lens protein (Scott and Bell, 1964), or the incorporation of amino acids into protein by germinating embryos in cottonseeds (Dure and Waters, 1965); in these cases the inference is that messenger RNA forms stable polysomes, which continue to serve as templates for protein synthesis during protracted periods. This also suggests that sRNA is conserved and used repeatedly for amino acid transfer during these periods, it being assumed that its *de novo* synthesis is arrested by actinomycin.

Most (85 percent) of the RNA in cells is in the form of ribosomal RNA (rRNA) combined with protein in the ribosomes, and much of the nonribosomal RNA is present as small molecules of transfer RNA (sRNA). These two forms, rRNA and sRNA, are comparatively stable, while messenger RNA exhibits varying degrees of instability. The experiments described above enabled "DNA-like RNA" to be isolated from bacteriophage-infected *E. coli* and furnished procedures for searching for sites on *E. coli* DNA which would complement with *E. coli* ribosomal RNA and sRNA.

One of the problems involved the fact that the larger of the two components of ribosomal RNA (the 23S component) has a molecular weight of about 10^6, so that even if it could be bound to *E. coli* DNA, the hybrid complex would involve only 0.02 percent of the total DNA of the *E. coli* chromosome. This problem was attacked by using labeled ribosomal RNA of high specific activity (Yankofsky and Spiegelman, 1962*b*). A further procedure took advantage of the fact that the RNA in hybrids is resistant to endonucleases. This enabled single-stranded unhybridized RNA to be removed by enzymatic digestion. It was found possible to prepare a hybrid between ribosomal RNA and homologous DNA which had RNA of the correct base composition and size as anticipated from the known base analysis and molecular weight of ribosomal RNA. In contrast, hybridization between 23S ribosomal RNA from *E. coli* and nonhomologous denatured DNA from T2 or T5 bacteriophage did not take place.

Subsequent experiments indicated that a complementary region also existed for the smaller rRNA component, 16S ribosomal RNA, and that this site was distinct from the site which complemented with 23S ribosomal RNA. This indicated that 16S is not a dimer of 23S even though the sedimentation rates indicate that 23S RNA has about twice the molecular weight of 16S sRNA.

Goodman and Rich (1962) investigated the origin of sRNA by using techniques similar to those that had been employed by Spiegelman and coworkers (Hall and Spiegelman, 1961; Yankofsky and Spiegelman, 1962b) in their studies of the genesis of mRNA and rRNA. The sRNA was labeled with P^{32} and was obtained from *E. coli*. Hybrid chains were formed when *E. coli* DNA and sRNA were heated to 70 C for 2 hours in 0.25 M NaCl, in 0.015 M sodium citrate buffer, pH 7.4, and were then slowly cooled during 15 hours to room temperature. About 0.1 µg of sRNA was hybridized by 40 µg of DNA. No annealing took place between *E. coli* sRNA and T2 viral DNA or salmon sperm DNA. The percentage of annealing between *E. coli* sRNA and various bacterial DNAs, expressed in terms of annealing by *E. coli* DNA = 100, showed interspecific relationships as expressed in Fig. 2-5. It is noteworthy that the tendency to anneal was related to the bacterial family classification rather than to the (A + T)/(G + C) ratio; *Brucella abortus* DNA showed less annealing with *E. coli* sRNA than did *Proteus vulgaris* DNA, in spite of the similarity in base ratios between *E. coli* and *Br. abortus*. Genetic exchanges have been shown to take place between *E. coli* and *Pr. vulgaris*.

The proportion of *E. coli* DNA (mol wt 4×10^9) hybridized by sRNA (mol wt 2.55×10^3) was about 0.025 percent. From this it was calculated that there are 40 sRNA sites in the *E. coli* genome. This is lower than the number 64, which would be needed if each triplet code corresponds to an anticodon (Table 2-16).

Similar experiments with similar results were reported by Giaconomi and Spiegelman (1962). Hybridization between *E. coli* sRNA and DNA indicated that sRNA saturated the DNA at values of 0.019 and 0.023 percent in separate experiments, equivalent to about 48 genetic sites in DNA for the formation of sRNA molecules. It must be assumed in such calculations, of course, that there is only one genetic site for each sRNA molecule. Spiegelman (1965) reported the existence of a multiplicity of genetic sites for ribosomal RNA; the number was greater in the most complex species of organisms that contained larger amounts of DNA per cell. Giaconomi and Spiegelman (1962) also noticed that sRNA from *Bacillus megaterium* would not hybridize with *E. coli* DNA or with *Pseudomonas aeruginosa* DNA. A further refinement in establishing the origin of rRNA was made by Oishi and Sueoka (1965), who succeeded in mapping its genetic locus on the chromosome of *B. subtilis*.

The interspecific relationships between sRNA molecules may be regarded as analogous to the corresponding situation in the cytochromes or the hemoglobins (Chapters 5 and 6) in which evolution has

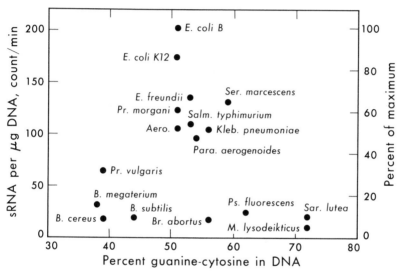

Fig. 2-5. Annealing of *E. coli* melted sRNA with DNA from various bacteria as related to GC content of DNA.

led to differences in nonfunctional regions of the primary structure while conserving the configuration of functional regions. The extent of the differences increases in proportion to the taxonomic separation of species, which in turn reflects an increasing time of separation from a common archetype.

An sRNA molecule has *functional sites* for bringing a certain amino acid to the correct locations in all the proteins of a species of living organism, as specified by a coding triplet in all the messenger RNA molecules of this species of living organism. These functional sites include the *recognition site*, for the amino acid as combined with the activating enzyme, and the *coding site*, which complements a specific triplet on messenger RNA. Mutational base substitutions that change or abolish the function of either of these two sites presumably will be lethal. A third functional site may be responsible for binding sRNA molecules to ribosomes as described below. The intermediate regions of the RNA strand, however, can undergo certain changes that may not interfere with function. The number of these changes will increase with the passage of time as the result of isolation brought about by speciation. These changes will give rise to progressively increasing differences between the sRNAs of different species which will be revealed by annealing experiments. This was shown by Goodman and Rich (1962). The changes reflect the sum of mutations in all the genes that code for the sRNA molecules.

Each species of organism will have its own family of 20 activating enzymes that participate in the reaction that couples an amino acid to sRNA. Each of these enzymes has evolved simultaneously with its *synonymous* sRNAs for the same amino acid to preserve a working relationship between the active site of the enzyme and the sequence of bases that the enzyme recognizes and combines with in the sRNA. We shall presume that in any given species the sRNAs for the same amino acid may differ with respect to their coding sites but not with respect to their amino acid recognition sites.

In terms of this model, the activating enzymes from yeast will be only partially effective in the aminoacylation reaction with sRNAs prepared from *E. coli* or in rabbit liver, as noted by various investigators (Berg et al., 1961; Benzer and Weisblum, 1961). This again reflects the divergence of species at the chemical level due to the time lapse associated with separate evolutionary pathways.

The sRNAs are remarkable for the presence of a large proportion of "unusual" bases distributed along their strands. The sRNAs vary among themselves with respect to the number and location of these bases, and these are apparently produced by enzymatic modification of the

Table 2-5. Unusual Bases or Ribosides in sRNA

Compound	Highest Level Found in sRNA,[a] mol/1,000 mol	Prominent Sources
1-methyladenine	...	Pig liver
2-methyladenine	6	S-180 ascites
6-methyladenine	20	Rat liver
6-dimethyladenine	5	S-180 ascites
1-methylguanine	10	Rat liver, rabbit liver
2-methylguanine	31	Mouse adenosarcoma
2-dimethylguanine	8	Mouse adenosarcoma
7-methylguanine	...	Pig liver
1-ribosylguanine	...	Yeast
3-methylcytidine	0.2	Yeast
5-methylcytidine	25	Rat liver
3-methyluridine	0.2[b]	Human liver
5-methyluridine	17.5[b]	E. coli
5-ribosyluracil	62.5	Rat liver
1,5-diribosyluracil	...[b]	Yeast
Inosine	3.3	Yeast
1-methylinosine	0.45	Yeast
6-(N-formyl-aminoacyl)adenine	...	Yeast[c]
2'-0-methyladenosine	2.6[a]	Sheep heart
2'-0-methylcytidine	3.4	Human liver
2'-0-methylguanosine	3.8[b]	Sheep heart
2'-0-methyluridine	3.1[b]	Sheep liver
2'-0-methylpseudouridine	...[b]	Sheep liver
5,6-dihydrouracil	3	Yeast[d]
4-thiouridylic acid	...	E. coli[e]
2-thiouridine	...	E. coli[f]

[a] sRNA except as noted by [b].
[b] Type of RNA not specified.
[c] Hall and Chheda (1965).
[d] Zamir, Holley, and Marquisee (1965).
[e] Lipsett (1965).
[f] Carbon, Hung, and Jones (1965).

Source: Compiled by R. H. Hall (personal communication) from publications of various authors.

four usual bases, A, C, G, and U, in the immature sRNA molecule. A list of the identified unusual bases in RNA is in Table 2-5, and their chemical formulas are in Fig. 2-6. The conclusion is that they are formed by modifications of A, C, G, and U in the chains of RNA by the action of enzymes which show marked species-specificity and

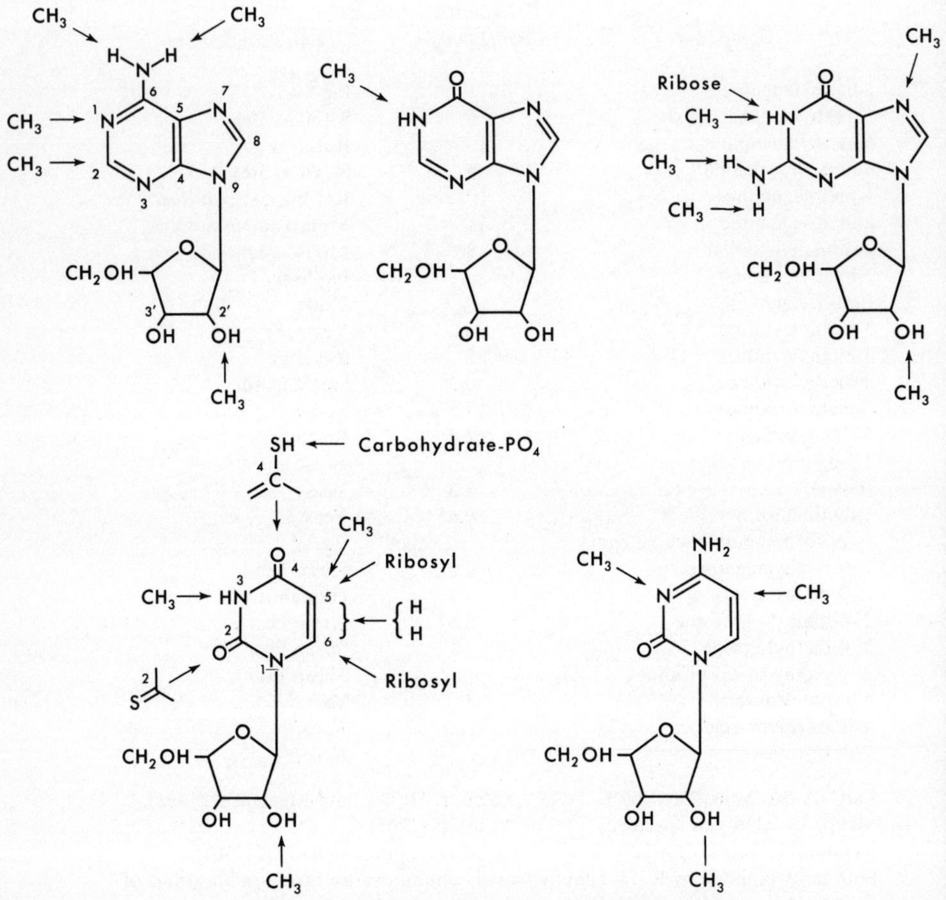

Fig. 2-6. Showing the attachment of special groups to the "unusual" bases in transfer RNA molecules as summarized in Table 2-5.

marked strain-specificity. In the case of RNA, methylating enzymes (Srinavasan and Borek, 1963; Gold, Hurwitz, and Anders, 1963; Svensson et al., 1963) are conspicuous among these enzymes, and Borek and his coworkers have shown that S-adenosylmethionine is the source of methyl groups. The enzymes are also substrate-specific for the bases and to some extent for the type of RNA. Hurwitz et al. (1965) studied a group of enzymes that catalyze the methylation of ribosomal RNA and found that only a small number of sites, between 0.1 and 0.2 percent of the total nucleotides, are methylated. These enzymes also catalyzed the methylation of sRNA, but a second group of enzymes which methylated sRNA did not act on rRNA. Unusual bases are present in small amounts in DNA as well as in RNA (Yu and Allen, 1959; Dunn, 1959).

The sRNAs are the most conspicuous among the nucleic acids for their content of unusual bases, prominent among which is 5-ribosyluracil (pseudouridine, ψ). Apparently, each sRNA undergoes a distinctive enzymatic modification that characterizes it for its specific function. Zamir, Holley, and Marquisee (1965) reported, however, that all sRNAs so far examined contain the same unique base sequence GpTpψpCpGp. This grouping may be concerned with attachment to ribosomes or perhaps with finding a specific site for the pyrophosphorylase that attaches the terminal CpCpA group to all sRNAs. It contains the unusual nucleosides thymine riboside, formed by S-adenosylmethionine-supplied methylation of uridine, and pseudouridine, formed by ribosidation of uridine at the 5 position of uracil. The sequence is isolated as the tetranucleotide TpψpCpGp from T1-ribonuclease treatment of sRNA. The other unusual bases appear to be scattered in various patterns through the different sRNA molecules.

It was found (Mandel and Borek, 1963) that the methylated bases are not necessarily needed for the transfer function of sRNA, since methionine starvation of a methionine-requiring mutant of *E. coli* led to the production of sRNA which contained abnormally low amounts of methylated bases. The sRNA produced under these conditions could be acylated with amino acids in the presence of appropriate activating enzymes (Starr, 1963). A leucyl-sRNA formed under these conditions was separated from other species of sRNA (Lazzarini and Peterkofsky, 1965). The question is open as to whether the unusual bases confer specific properties upon sRNAs that are of importance in protein synthesis or whether the unusual bases may help to protect the sRNAs against enzymatic degradation by nucleases.

The enzymatic methylation of DNA, like that of RNA, was found to be carried out by enzymes that are species- and strain-specific (Gold and Hurwitz, 1964). An enzyme was found to be present in *E. coli* which produced 5-methylcytosine and 6-methylaminopurine in various DNAs. The number and location of sites that were methylated were characteristic for each DNA.

The Amino Acid Code

The first definite information on the composition of the amino acid code was obtained by experiments in cell-free systems, prepared from bacteria, as first used successfully by Nirenberg and Matthaei (1961). The principle of this system is as follows: The protein-synthesizing components of the broken cells are separated from the bulk of the cell contents, including the cell walls and nuclei, by gentle centrifugation. During these manipulations, the natural messenger RNA tends to become broken down by enzymatic processes. An artificial messenger is now introduced which consists of a synthetic polyribonucleotide preparation of known base composition and varying strand lengths. Simultaneously, a single radioactive amino acid and a mixture of "cold" amino acids are added, and the system is incubated for a short time. Polypeptide synthesis is detected when a reagent, such as trichloracetic acid or tungstic acid, that precipitates polypeptides but not free amino acids is added. The radioactivity of the precipitate indicates the extent to which the amino acid has been incorporated into polypeptide linkage. The existence of such linkages may be confirmed by treatment of the precipitate with proteolytic enzymes (Kaziro, Grossman, and Ochoa, 1963).

It was found by Nirenberg and Matthaei (1961) that a polyribonucleotide consisting of a single nucleotide, uridylic acid, produced polyphenylalanine in the cell-free system that they prepared from *E.*

coli. This celebrated experiment was the key to a new biochemical understanding of genetics. The success of the experiment was aided by two important facts, first, polyuridylic acid forms strands that do not become tangled within themselves or with each other. The strands therefore can readily become attached to ribosomes to form *polysomes*, the multiple units that function in protein synthesis. Second, polyphenylalanine chains, even of short length, are much less soluble than the free amino acid phenylalanine and are readily precipitated by trichloracetic acid.

Experiments of the same type were carried out with other polyribonucleotides. They were prepared with the enzyme polynucleotide phosphorylase (Heppel, Ortiz, and Ochoa, 1957), which can be obtained from bacteria such as *Azotobacter vinelandii* or *M. lysodeikticus*. The enzyme differs from RNA polymerase in not requiring the presence of a DNA primer or template for the synthesis of RNA. Instead, it produces long polynucleotide strings with bases that are arranged randomly in frequencies that depend directly on the proportion of the different nucleoside diphosphates used as substrates in the reaction mixture. These are uridine, cytidine, adenosine, and guanosine diphosphates.

If a mixture of $5x$ parts of the letter U and x parts of the letter C is randomized into a long sequence, the resultant chain will contain three-letter groups in a following proportions: UUU, 125; UUC, UCU, and CUU, each 25; UCC, CUC, and CCU, each 5, and CCC, 1, subject, of course, to statistical variation. If the proportions of U and C are changed, the relative frequency of the U and C triplets will be altered in a predictable manner. It was found (Lengyel, Speyer, and Ochoa, 1961) that polyribonucleotides containing U and C would lead to the incorporation of four, and only four, amino acids—phe, leu, ser, and pro—into polypeptides in the cell-free system. Eight triplets or four doublets (UU, UC, CU, and CC) can be formed by U and C. The results, therefore, did not allow a decision to be made as to whether the amino acid code consisted of doublets or triplets.

Similar experiments with copolymers of U and A, however, produced incorporation of six amino acids—phe, tyr, ilu, leu, asN, and lys (Speyer et al., 1962). This result made it probable that the code consisted of triplets rather than doublets. It was subsequently shown that the results with polyribonucleotides containing U and C could be explained on the basis that phe, leu, ser, and pro were each coded by two of the eight possible triplets formed by U and C. The probability of a triplet code was increased by experiments with copolymers containing three different bases. It was found, for example, that a copolymer containing U, A, and G would incorporate methionine but that this amino acid was not incorporated by U + A, U + G, or A + G copolymers (Speyer et al., 1962).

Polyribonucleotides containing high proportions of G were difficult to use in the cell-free system. This difficulty was attributed to the tendency of G to form cross-linkages by hydrogen bonding that tangled the strands and prevented the polyribonucleotides from combining effectively with ribosomes. No such problems were encountered with A, C, and U, so that it was possible to assign coding functions to all 27 possible triplet permutations of A, C, and U. These were summarized by Speyer et al. (1962), as shown in Tables 2-6 to 2-11. These tables show the procedures and calculations used in this approach to the assignment of the RNA triplets in the amino acid code. Later it was shown that the codes for glutamic acid, glutamine, and lysine did not contain U and that the codes for asparagine were 2A,1C and 2A,1U but not 1A,1C,1U.

In a somewhat different approach, similar and concordant results for AC and CU were obtained by Nirenberg et al. (1963). They prepared copolymers containing varying proportions of AC and CU. These were used for amino acid incorporation in the cell-free system. The observed frequencies of incorporation of amino acids were plotted against the calculated frequency of the triplets in the copolymers. Examples of the results are shown in Fig. 2-7.

Various approaches are possible for assigning sequences to the triplets. One method is by experiments on amino acid incorporation with messengers of known base composition and sequence. The first experiments of this type were reported by Wahba et al. (1963): Polynucleotide phosphorylase can use short lengths of polyribonucleotides

Table 2-6. Amino Acid Incorporation in *E. coli* System with Various Polynucleotides[a]

Amino Acid	UC (5:1)		UC (4.3:5.7)		UC (1:5)		AC (5:1)		AC (1:5)		UAC (1:1:4)		UAC (6:1:1)	
	None	160 μg/ml	None	320 μg/ml	None	320 μg/ml	None	80 μg/ml	None	320 μg/ml	None	320 μg/ml	None	240 μg/ml
asN	...	...	...	...	...	...	47	1,146	27	98	65	553	...	...
glN	...	...	...	...	...	...	39	1,117	31	101	137	552	67	167
his	...	...	...	...	...	...	282	576	28	343	59	2,025	...	...
ilu	...	...	...	...	...	...	...	...	...	...	54	397	...	...
leu	74	3,368	49	4,100	69	659	...	...	...	...	...	...	...	...
lys	...	...	...	...	...	...	60	4,615	12	26	...	...	...	...
phe	49	14,860	96	4,516	79	159	...	...	...	...	157	360	173	3,432
pro	21	784	47	7,785	51	2,880	14	342	8	1,350	30	5,680	...	...
ser	46	3,540	41	4,035	44	586	...	...	...	...	...	...	...	...
thr	...	...	...	...	...	...	44	1,250	22	301	45	2,145	...	...
tyr	...	...	...	...	...	...	...	...	...	...	50	277	...	...

[a] μμmoles/mg ribosomal protein. The following precipitating reagents were used: 5 percent trichloroacetic acid in the experiments with poly-UC (5:1) and UAC (6:1:1); 20 percent trichoroacetic acid in those with poly-UC (4.3:5.7), UC (1:5), AC (1:5), and UAC (1:1:4); and 0.25 percent sodium tungstate in 5 percent trichloroacetic acid in those with poly-AC (5:1). The results presented in this table are average values of an experiment done with duplicates. Similar results were obtained in at least one other experiment.

The base ratios given in parentheses are the ratios of nucleoside diphosphates used in the preparation of the polymers except for poly-UC (4.3:5.7). In this case, the actually determined base ratio is given (the polymer was prepared with UDP and CDP in 1:1 ratio). The base ratios of poly-UC (5:1) and -AC (5:1) as actually determined were 4.7:1 and 4.9:1, respectively.

Source: Speyer et al. (1963).

Table 2-7. Data for Code Triplet Assignments (Copolymers of A and C)

Amino Acid	Calculated Triplet Frequency				Sum of Calculated Triplet Frequencies	Amino Acid Incorporation[a]
	3A	2A1C	1A2C	3C		
			Poly AC (5:1)			
asN	...	20	...	...	20	24.2
glN	...	20	...	...	20	23.7
his	...	...	4.0	...	4	6.5
lys	100	...	...	...	100	100
pro	...	...	4.0	0.8	4.8	7.2
thr	...	20	4.0	...	24	26.5
			Poly AC (1:5)			
asN	...	3.3	...	...	3.3	5.3
glN	...	3.3	...	...	3.3	5.2
his	...	...	16.7	...	16.7	23.4
lys	0.7	...	...	...	0.7	1.0
pro	...	...	16.7	83.3	100	100
thr	...	3.3	16.7	...	20	20.8

[a] These values are calculated from the data of Table 2-5. The net promotion of the incorporation of an amino acid by a certain polymer is given as percent of the net promotion of incorporation of the amino acid whose incorporation is promoted to the greatest extent. The amino acid in question is phe for U-rich polymers, pro for C-rich polymers, and lys for A-rich polymers.
Source: Speyer et al. (1963).

Table 2-8. Data for Code Triplet Assignments (Copolymers of U and C)

Amino Acid	Calculated Triplet Frequency				Sum of Calculated Triplet Frequencies	Amino Acid Incorporation[a]
	3U	2U1C	1U2C	3C		
			Poly UC (5:1)			
leu	...	16.7	3.3	...	20	22.2
phe	83.3	16.7	...	...	100	100
pro	...	...	3.3	0.7	4	5.1
ser	...	16.7	3.3	...	20	23.6
			Poly UC (4.7:5.3)			
leu	...	53	60	...	113	92
phe	47	53	...	...	100	100
pro	...	...	60	67	127	175
ser	...	53	60	...	113	90
			Poly UC (1:5)			
leu	...	3.3	16.7	...	20	20.8
phe	0.7	3.3	...	...	4	2.8
pro	...	...	16.7	83.3	100	100
ser	...	3.3	16.7	...	20	19.2

[a] See note to Table 2-6.
Source: Speyer et al. (1963).

Table 2-9. Data for Code Triplet Assignments (Copolymers of A and U)

Amino Acid	Calculated Triplet Frequency				Sum of Calculated Triplet Frequencies	Amino Acid Incorporation[a]
	3A	2A1U	1A2U	3U		
			Poly AU (5:1)			
asN	...	16.7	...	...	16.7	28
ilu	...	16.7	3.3	...	20	20
leu	...	...	3.3	...	3.3	3.4
lys	83.3	16.7	...	...	100	100
phe	...	...	...	0.8	0.8	0
tyr	...	...	3.3	...	3.3	3.2
			Poly AU (1:5)			
asN	...	4	...	...	4	6.6
ilu	...	4	20	...	24	20
leu	...	...	20	...	20	15
lys	0.8	4	...	...	4.8	3.1
phe	...	...	...	100	100	100
tyr	...	...	20	...	20	25

[a] See note to Table 2-6.
Source: Speyer et al. (1963).

Table 2-10. Data for Code Triplet Assignments, Poly UAC (1:1:4)

Amino Acid	3C	2C1A	2C1U	1C2A	1C1A1U	1C2U	2A1U	1A2U	3U	Sum of Calculated Triplet Frequencies	Amino Acid Incorporation[a]
asN	…	…	…	4.2	4.2	…	1.0	…	…	9.4	8.6
glN	…	…	…	4.2	4.2	…	…	…	…	8.4	6.8
his	…	16.7	…	…	4.2	…	…	…	…	20.9	34.8
ilu	…	…	…	…	4.2	…	1.0	1.0	…	6.2	6.1
phe	…	…	…	…	…	4.2	…	…	1.0	5.2	3.6
pro	66.7	16.7	16.7	…	…	…	…	…	…	100.0	100.0
thr	…	16.7	…	4.2	4.2	…	…	…	…	25.1	37.2
tyr	…	…	…	…	4.2	…	…	…	…	5.2	4.0

[a] See note to Table 2-6.
Source: Speyer et al. (1963).

as primers for elongation into longer strands. It is possible to prepare such primers consisting of AU and AAU by treating a copolymer of U and A (5:1) with pancreatic ribonuclease. The treatment produces Up, ApUp, and ApApUp, theoretically in the proportion of 25:5:1. The primers may be elongated into AUUUUU...U and AAUUUUU...U copolymers by incubation with uridine diphosphate

Table 2-11. Code Triplets Occurring in Copolymers of A, C, and U: Amino Acids Assigned to the Triplets, Summarizing Tables 2-6 to 2-10

Triplet Composition	No. of Triplets	Amino Acid
3A	1	lysine
2A1C	3	asparagine, glutamine, threonine
2A1U	3	asparagine, isoleucine, lysine[a]
1A2C	3	histidine, proline, threonine
1A1C1U	6	asparagine,[a] glutamine,[a] histidine, isoleucine, threonine, tyrosine
1A2U	3	isoleucine, leucine, tyrosine
3C	1	proline
2C1U	3	leucine, proline, serine
1C2U	3	leucine, phenylalanine, serine
3U	1	phenylalanine

[a] These assignments were subsequently shown to be erroneous.
Source: Speyer et al. (1963).

and polynucleotide phosphorylase. The experiments are difficult to carry out because of the presence of enzymes in the cell-free system that break down the polyribonucleotides. Small amounts of carboxyl-terminal tyrosine were detected in the polyphenylalanine fraction, and the evidence was thought to indicate the sequence of the 1A,2U triplet for tyrosine as AUU. Subsequently (p. 58), the sequence for tyrosine was found to be UAU, and the results by Wahba et al. (1963) were not confirmed.

Mutations and the Code

We can determine the primary structure of proteins by dividing them into small polypeptides with suitable enzymes and then measuring the amino acid sequence in each polypeptide. This procedure enables

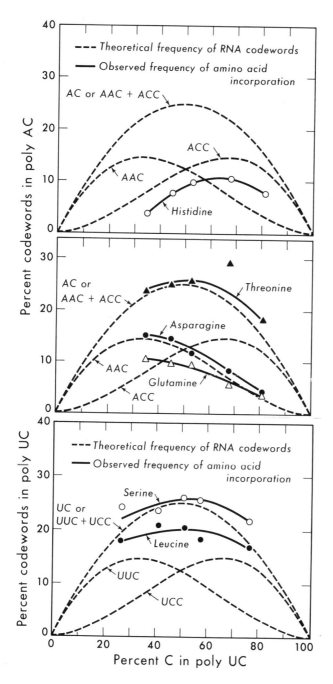

Fig. 2-7. Incorporation of amino acids in cell-free system compared with theoretical triplet content of synthetic polyribonucleotides (from Nirenberg et al., 1963). The results indicate the following code assignments: his, thr, A + 2C; thr, asN, glN, 2A + C; ser, leu, 2U + C and U + 2C.

mutational changes of a single amino acid at a definite locus to be detected. The earliest and most important example was the discovery of the substitution of valine for glutamic acid at position 6 in the β chain of human hemoglobin. This single-amino-acid mutation in a polypeptide chain of 146 amino acids was identified by Ingram, who attributed it to a single-base change in the DNA of the gene responsible for hemoglobin (Ingram, 1958). Other single-amino-acid mutations have since been discovered—some in human hemoglobin, some produced by chemical treatment of tobacco mosaic virus (TMV) or its RNA component, and some in the A protein of tryptophan synthetase, produced by treating *E. coli* with ultraviolet light or chemicals (Table 4-5). All of them correspond to single-base changes in coding triplets (Table 4-5.) Certain deductions regarding the sequences of bases in the triplets may be made from the mutations if the following assumptions are made:

1. Single-amino-acid mutations are caused by single-base changes in the coding triplets.

2. When two or more coding triplets for the same amino acid contain the same pair of bases, or *shared doublet* (Wahba et al., 1963; Jukes, 1963*b*), this pair usually occupies the same relative position within the triplets; for example, if one code for glycine is GGU, two others are GGA and GGC.

An argument, which has been advanced by Wittmann and Wittmann-Liebold (1963) is that such an arrangement of the bases lessens the number of amino-acid changes that are produced by base changes and hence stabilizes the proteins in a living organism.

3. The third argument is based on the feeling that nature is often orderly and that there is less distortion of a transfer enzyme for glycine if three triplets for this amino acid can be written as GGA, GGC, and GGU than if they are written as, for example, AGG, GGC, and GUG.

Binding between sRNAs and Trinucleotides

It was found by Nirenberg and Leder (1964; Leder and Nirenberg, 1964*a*) that specific interactions occurred between trinucleotides, ribosomes, and aminoacyl sRNA in the following manner: Ribosomes were mixed with an oligonucleotide at pH 7.2 in the presence of Mg^{++} and K^+ at 0 C. A C^{14}-tagged amino acid was attached to its specific sRNA, or sRNAs, by charging a mixture of sRNAs with a mixture of the C^{14} amino acid and the other 19 cold amino acids. The mixture of sRNAs and ribosomes was incubated for 20 minutes at 24 C and poured through a Millipore filter, which retained the ribosomes and

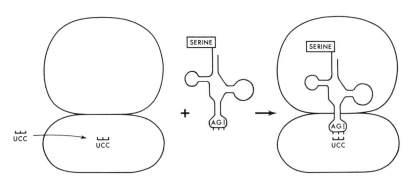

Fig. 2-8. Diagrammatic representation of the attachment of the trinucleotide UCC to a ribosome so that it binds only seryl-sRNA.

any bound sRNA. Binding of C^{14}-phe-sRNA was produced by poly-U, of C^{14}-lys-sRNA by poly-A, and of C^{14}-pro-sRNA by poly-C. The trinucleotides pUpUpU or, less effectively, UpUpU induced phe-sRNA binding, but UpUpUp was inactive, as were the dinucleotides pUpU, UpUp, and UpU. The procedure is shown diagrammatically in Fig. 2-8. The results could be interpreted as indicating that a trinucleotide with a terminal 3'OH group could become bound to the site on the ribosome that is normally occupied during protein synthesis by a messenger RNA coding triplet that attaches to the complementary coding triplet sequence of an aminoacyl sRNA. The trinucleotide simulates a short piece of messenger RNA and binds the specific charged sRNA to the ribosome. Additional results (Leder and Nirenberg, 1964a) showed that the binding of valyl-sRNA to ribosomes was brought about by GpUpU but not by UpGpU or UpUpG. No aminoacyl-sRNA's other than valyl-sRNA were bound to ribosomes by GpUpU. Leucyl-sRNA was not bound by any of these three trinucleotides in spite of the fact that 2U,IG is a code for leu.

In their next publication Leder and Nirenberg (1964b) found that UpUpG bound a special preparation of leucyl-sRNA, fraction II-G of Weisblum, Benzer, and Holley (1962). Cysteinyl-sRNA was bound by UpGpU. Presumably, the negative results obtained for leucine earlier with UpUpG were due to the absence of the specific sRNAs containing the appropriate coding sequence, $\overleftarrow{ApApC}$ or $\overleftarrow{ApApU}$, needed to complement the messenger coding triplet $\overrightarrow{UpUpG}$.

The assignments of GUU to val, UGU to cys, and UUG to leu signaled a rapid advance in knowledge of the genetic code. These assignments may be used to order the sequences in other coding triplets for amino acids that are related to val, cys, and leu by mutations. For example, a series of mutations connect val and cys as follows: val/gly/arg/his/tyr/cys, which could be written as GUU/GGU/CGU/CAU/UAU/UGU. Other series are val/ilu/met/leu, which could be GUU/AUU/AUG/UUG, and val/ala/asp/asN/ser/gly could be GUU/GCU/GAU/AAU/AGU/GGU. Most of these trinucleotides were soon tested by Nirenberg's group by means of the ribosome-sRNA binding procedure. The solution to the sequence problem was at hand. It differed from a proposal in earlier publications (Speyer et al., 1962; Wahba et al., 1963; Jukes, 1963a), in which a code and sequence of AUU were assigned to tyr and were used to order the other triplets, but most of these could be corrected by transposing the bases from the third position to the first position, e.g., tyr, UAU.

Trinucleotides containing C and U were examined by Bernfield and Nirenberg (1965), again with respect to the binding of C^{14}-aminoacyl-sRNA to ribosomes. The trinucleoside diphosphates UpUpU, UpUpC, UpCpU, CpUpU, CpCpU, CpUpC, UpCpC, and CpCpC were prepared by the enzymatic procedure of Heppel, Whitfield, and Markham (1955), in which pyrimidine-2',3'-cyclic nucleotides are attached enzymatically to dinucleoside phosphates. The base ratio and sequences were determined by enzymatic degradation. The test of sRNA binding was carried out in the presence of washed ribosomes, 0.05 M KCl, and 0.02 or 0.03 M magnesium acetate in a pH 7.2 buffer.

The binding of phe-sRNA by polyU was much greater than by any of the trinucleotides, but pUpUpU, UpUpC, and UpUpU bound phe-sRNA in significant quantities. The other trinucleotides did not bind phe-sRNA nor did any of the dinucleotides containing U and C. Binding of ser-sRNA at a level of slightly less than twice the background was obtained with UpCpU and UpCpC. Leu-sRNA was not bound by any of the trinucleotides containing C and U. The authors concluded, however, that CUU and CUC might be sequences for leu because poly-UC was active in this respect. They also assigned a tentative sequence of CCU = pro.

Better evidence for assigning CUC as a code for leu was obtained by Nishimura et al. (1965). These investigators used polynucleotides of alternating sequence, prepared by enzymatic treatment of DNA templates of the appropriate complementary content, using RNA polymerase. The composition of the copolymers was established by nearest-neighbor analyses. A polynucleotide consisting of UCUCUC... was found to serve as a template for the production of a polypeptide in which leu alternated with ser, leading to the conclusion that CUC is a code for leu and UCU for ser. By similar procedures, the polynucleotide UGUGUG... was found to incorporate cys and val, confirming the finding that UGU = cys and indicating that GUG is a second code for val.

Nishimura and coworkers also prepared the copolymer

AAGAAGAAG...,

which contains the three sequences GAA, AGA, and AAG, depending upon where the "reading" of the message starts in the sequence along the RNA strand. The copolymer gave excellent incorporation of lysine into polylysine in an amino-acid-incorporating system, being superior to poly-A for this purpose. It also directed the synthesis of polyarginine and polyglutamic acid. The action of exonucleases could change the starting point of this copolymer in the system. It was then found, by means of trinucleotides in a system similar to that employed by Leder

Table 2-12. The Amino Acid Code

Triplet	Amino Acid[a]	Source	Triplet	Amino Acid[a]	Source
UUU	phe	1	AUU	ilu	1, 3, 8
UUC	phe	1, 8	AUC	ilu	1, 8
UUA	leu	7	AUA	*ilu*	5
UUG	leu	1, 8	AUG	met	1, 3, 7, 8
UCU	ser	1, 2, 8	ACU	thr	1, 8
UCC	ser	1, 8	ACC	thr	1, 8
UCA	ser	7, 8	ACA	thr	1, 2, 8
UCG	ser	1, 8	ACG	thr	1, 8
UAU	tyr	1, 3, 8	AAU	asN	1, 7, 8
UAC	tyr	1, 3, 8	AAC	asN	1, 4, 8
UAA	*Gap*	6, 9	AAA	lys	1, 8
UAG	*Gap*	6, 9	AAG	lys	1, 2, 3, 8
UGU	cys	1, 2, 8	AGU	ser	1, 7, 8
UGC	cys	1, 8	AGC	ser	1, 8
UGA	(?)	(?)	AGA	arg	2, 8
UGG	try	6, 8	AGG	*arg*	5
CUU	leu	1, 3, 7	GUU	val	1, 8
CUC	leu	1, 2, 8	GUC	val	1, 7, 8
CUA	*leu*	5	GUA	val	1, 8
CUG	leu	1, 8	GUG	val	1, 8
CCU	pro	1, 8	GCU	ala	1, 8
CCC	pro	1, 8	GCC	ala	1, 8
CCA	pro	1, 7, 8	GCA	ala	1, 8
CCG	pro	8	GCG	ala	1, 8
CAU	his	1, 7, 8	GAU	asp	1, 8
CAC	his	1, 2, 7, 8	GAC	asp	1, 8
CAA	glN	1, 8	GAA	glu	1, 2, 8
CAG	glN	1, 8	GAG	glu	1, 2, 8
CGU	arg	1, 8	GGU	gly	1, 8
CGC	arg	1, 8	GGC	gly	8
CGA	arg	1, 8	GGA	gly	8
CGG	arg	1, 8	GGG	gly	8

[a] All assignments except those italicized are based on direct experimental findings.

Sources:
1. Trupin et al. (1965); Nirenberg et al. (1965); Brimacombe et al. (1965).
2. Nishimura et al. (1965).
3. Thach, Sundararajan, and Doty (1965).
4. Smith et al. (1965).
5. Deduced from unordered triplet assignments, homology, and mutations.
6. Brenner, Stretton, and Kaplan (1965); Weigert and Garen (1965).
7. Terzaghi et al. (1965); see also Chapter 4.
8. Söll et al. (1965).
9. Takanami and Yan (1965).

and Nirenberg, that ApApA and ApApG both bound unfractionated lysyl sRNA; ApApA was superior to ApApG in this respect. GpApA and ApGpA did not bind lysyl sRNA, but GpApA bound glutamyl sRNA in the presence of 0.01 M Mg^{++}; ApGpA and ApApG were inactive. The conclusion was that GpApA was a code for glu.

Continuing their investigations with ribonucleotide copolymers, the New York University group (Smith et al., 1965) prepared A + C (25:1) and repeated their previous finding that lys, asN, and thr were joined into polypeptide sequences by its use in a cell-free amino-acid-incorporating system. In this case *E. coli* ribosomes were used in combination with supernatant fraction prepared from *Lactobacillus arabinosus*, which is low in nucleases. Another aliquot of the poly-A + C was then treated with pancreatic ribonuclease to break the CpA linkages at Cp, and the terminal 3'-phosphate groups were removed enzymatically, so that the preparation should then consist of pApAp ... $pApApC_{OH}$ copolymers. It was found to incorporate lys and small amounts of asN but no thr, indicating that AAC is a code for asN. Furthermore, the asN thus incorporated was shown to occupy the carboxy-terminal position in a peptide chain, thus indicating that the base sequence in the polyribonucleotide was translated in a "left-to right" direction. The argument for this is as follows: Polypeptide chains are known to be synthesized starting at the amino-terminal end and finishing with the amino acid having a free terminal carboxyl group (p. 23). The polynucleotide ended in ... ApApC. Therefore, the genetic message was translated from left to right.

Further progress on the coding assignments of triplets was described in reports on the template activities of ordered trinucleotides tested by means of the ribosomal-sRNA binding procedure (Trupin et al., 1965; Nirenberg et al., 1965; Brimacombe et al., 1965; Söll et al., 1965). These findings served to dispose of most remaining uncertainties in the matter of the sequential arrangements of bases in the messenger RNA (Trupin et al., 1965) coding triplets, and the amino acid code is in Table 2-12. A condensation of Table 2-12 is shown in Table 2-13.

Simultaneously, other announcements appeared regarding the sequences in certain coding triplets as a result of experiments with ordered copolymers of ribonucleotides. Söll et al. (1965), continuing the experiments described on p. 57, found that the incorporation of amino acids into polypeptides was produced by copolymers as follows: $(pApApG)_n$, lys, glu, arg; $(pApG)_n$, arg, glu; $(pApC)_n$, his, thr. These results indicated the following codes in addition to those established by their previous findings; AGA, arg; GAG, glu; ACA, thr; CAC,

Table 2-13. The Amino Acid Code in Abbreviated Form

UUb phenylalanine	CUd leucine	AUb isoleucine	GUd valine
		AUA isoleucine	
UUe leucine		AUG methionine	
UCd serine	CCd proline	ACd threonine	GCd alanine
UAb tyrosine	CAb histidine	AAb asparagine	GAb aspartic acid
UAe gaps	CAe glutamine	AAc lysine	GAe glutamic acid
UGb cysteine	CGd arginine	AGb serine	GGd glycine
UGe tryptophan		AGe arginine	

U = uracil, C = cytosine, A = adenine, G = guanine
b = U, C; d = U, C, A, G; e = A, G
Sources: See Table 2-12.

and CAC, his. Thach, Sundararajan, and Doty (1965) used short oligonucleotides to stimulate the specific binding of amino-acyl sRNAs to ribosomes. Block oligonucleotides A_4U_5 and UAU_7 bound ilu; CU_3 and CU_4 bound leu, while A_4U_6 incorporated lys, ilu, and phe into peptides. These and other results indicated codes of AUU, ilu; CUU, leu; UAU and UAC, tyr; AAG, lys; and AUG, met. AUA did not bind methionine sRNA. These results and those of Terzaghi et al. (1965) are all included in Table 2-12. This table eliminates the following assignments that were proposed in earlier studies (p. 52): 1A,1C,1U, asN, glN; 1A,1G,1U, glu; 1A,2G, glN; and 2A,1U, lys.

The identity of the "nonsense," or "amber" and "ochre," triplets, which terminate the transcription of polypeptide chains, was revealed as UAG and UAA by studies with alkaline phosphatase of *E. coli* (Weigert and Garen, 1965) and r-II mutants of the A cistron of the T4 bacteriophage of *E. coli* (Brenner, Stretton, and Kaplan, 1965). In the first-mentioned investigation, phosphatase-positive revertants were produced by treating phosphatase-negative "nonsense" mutants with a mutagen. The amino acid replacements that restored function to the nonsense mutant were identified by fingerprinting the triplet peptides of the revertant alkaline phosphatases. The replacements were glN, glu, leu, lys (or arg), ser, try, and tyr. This enables the chain-terminating triplet to be identified as UAe (e = A, G), giving all seven possible single-base replacements CAe, GAe, UUe, AAe, UCe, UGe, and UAc (c = U, C) and excluding arg. The second investigation led to the same conclusions and is described in Chapter 4.

Synthetic polyribonucleotides were used by Takanami and Yan (1965) as sources of mRNA in a cell-free amino-acid-incorporating system to study the release of polypeptide chains from ribosomal attachment. The use of the following polynucleotides resulted in most (about 90 percent) of the polypeptide material remaining bound to the ribosomes: UC, UI, UCI, AC, AIC. I was used instead of G because of its similar properties in complementing with C and the greater ease of preparing and using polymers containing I than polymers containing G. In contrast the use of UA, UAI, UAC, and UAIC led to the release of substantial percentages of the incorporated amino acids into the supernatant solution. Evidently, the presence of both U and A in the polynucleotide was necessary for detaching the polypeptide chain from ribosomal binding. The binding is due to linkage of the chain through an ester bond to an sRNA molecule which is held by the 50s ribosomal particle. It was also noted that the amount of chain release by UA copolymers increased as the molar percentage of A was increased from 11.5 to 21 percent and that a UAI copolymer was more effective than UAC copolymer which had a similar quantitative composition except that C replaced I. The conclusion was that 2A,1U and 1A, 1U,1I triplets were instrumental in the termination and release of polypeptide chains. This supports the proposal by Brenner et al. (1965) that UAA and UAG are chain-terminating triplets.

The incorporation of serine by the UAI copolymer was studied by Takanami and Yan (1965) in two different cell-free systems obtained, respectively, from wild type (B) *E. coli* and from a suppressor mutant CR63. About 40 percent more serine was incorporated by CR63 than

Table 2-14. Incorporation of Amino Acids by Poly-UAI, UCI, UAC, and CAI in *E. Coli* B and CR63 Cell-Free Systems[a]

Amino Acid	UAI, percent of phe			UCI, percent of phe			UAC, percent of phe			CAI, percent of pro		
	Calculated	Found B	Found CR63	Calculated	Found B	Found CR63	Calculated	Found B	Found CR63	Calculated	Found B	Found CR63
phe	(100)	(100)	(100)	(100)	(100)	(100)	(100)	(100)	(100)	0	0	0
leu	79	87	86	210	219	195	106	117	108	0	0	0
ser	15	19	30	176	208	194	89	93	92	6	16	17
ileu	41	48	45	0	22	26	37	47	50	0	±	±
val	84	97	92	108	128	118	0	0	0	0	0	0
ala	0	0	0	140	133	120	0	0	0	43	36	36
gly	40	31	34	86	92	87	0	0	0	19	12	13
thr	0	0	0	0	38	35	28	33	31	22	29	27
his	0	0	0	0	0	0	24	19	19	13	13	15
lys	8	14	13	0	0	0	2	2	5.0	2	2.5	3.1
arg	12	15	14	140	202	186	0	5.0	5.0	47	55	57
asp	15	7.5	7.5	0	0	0	0	±	±	6	+	+
glu	12	9.3	9.1	0	0	0	0	+	+	4	5.1	6.9
pro	0	0	0	228	185	176	66	59	69	(100)	(100)	(100)
tyr	31	25	26	0	0	0	36	34	36	0	0	0
met	15	24	25	0	±	±	0	±	±	0	0	0
try	22	18	17	28	26	26	0	0	0	0	0	0
cySH	47	44	49	80	73	55	0	0	0	0	0	0

[a] Data for glu and asp are regarded as inaccurate, because glu ⇌ gluN and asp ⇌ aspN changes may occur during incubation. The symbol ± indicates less than 2× the amount of incorporation without copolymer; +, more than 2×. In *E. coli* B, incorporation of phe was 1.9 mμ moles by UAI, 1.1 mμ moles by UCI, and 2.2 mμ moles by UAC, and that of pro was 1.7 mμ moles by CAI. The calculated values are from the following assignments: phe, UUb; leu, UUe, CUd; ser, UCd, AGb; ileu, AUb, AUA; val, GUd; ala, GCd; gly, GGd; thr, ACd; his, CAb; lys, AAe; arg, CGd, AGe; asp, GAb; glu, GAe; pro, CCd; tyr, UAb; met, AUG; try, UGG; cySH, UGb (b = U,C; d = A,G,U,C; e = A,G). Italicized values indicate unusual or unanticipated increases in incorporation.

Source: Takanami and Yan, 1965.

by B. Twelve other amino acids showed no such differences. All comparisons were on the basis of phe = 100. The results are in Table 2-14.

The conclusion is reached that the code shows an orderly pattern. This was foreshadowed by the "shared-doublet" concept (p. 54), which stated that in many cases a common base pair occupied the same portion in two or more of the synonymous coding triplets for an amino acid. The principle of homology in base sequences of synonymous triplets is now established (Table 2-12). It appears from Table 2-13 that the bases that were formerly termed shared doublets correspond to the bases in positions 1 and 2 of a triplet and that U and C can be interchanged in position 3 without altering the coding properties of a triplet.

The 64 triplets in the amino acid code included 16 pairs of triplets, each consisting of the general formula XYC and XYU, where XY is any of the permutations of two members of the group A, C, G, and U. These 16 pairs code, respectively, for the amino acids asN, ser, thr, ilu, asp, gly, ala, val, his, arg, pro, leu, tyr, cys, and phe. The other 16 pairs of triplets in the table each consist of the general formula XYA and XYG. In the cases of XY = AC, GG, GC, GU, CG, CC, CU, and UC, the "quartet" of triplets represented by XYA, XYG, XYC, and XYU code in each case for a single amino acid. The amino acids in this group are thr, gly, ala, val, arg, pro, leu, and ser. When XY = AA, AG, AU, GA, CA, UA, UG or UU, this is not the case; indeed, it is arithmetically obvious that there are only 16 quartets in the 64 triplets and 20 amino acids are coded. The assignments for some of the triplets are difficult to decide by the trinucleotide-ribosome-sRNA procedure if the mixture of sRNAs is deficient in any sRNA molecule needed for complementation with a specific trinucleotide being used in the test. Thus, a negative test with a trinucleotide, such as UpGpA, may mean either that the appropriate sRNA was absent or that the triplet had no coding function. The point can be resolved with additional study by means of other approaches. UGA is not yet assigned.

The pattern of regularity in the code in Table 2-13 emphasizes that the bases in the first two positions in the triplet are usually definitive and the third is variable. This interesting observation leads to speculations regarding the evolution of the genetic code and of the amino acids themselves.

A widely discussed proposal for the origin of life includes the speculation that the primitive terrestrial atmosphere contained no free oxygen but was rich in hydrocarbons and that nitrogen first appeared on the Earth's surface not as N_2 but in the form of ammonia (Oparin, 1964, p. 52). When the Earth cooled off sufficiently to permit the formation of the oceans, according to this hypothesis, organic compounds were produced in solution or were formed in the reducing atmosphere as a result of energy supported by light and electrical storms. This gave rise to the "primeval nutritive soup" (*ibid.*, p. 65).

These speculations received a strong stimulus when Miller (1955) produced traces of glycine, alanine, aspartic acid, and glutamic acid from ammonia, methane, and water by subjecting them to electrical discharge. The work in this field was reviewed by Harada and Fox (1965), who heated methane, ammonia, and water at 900–1100 C in the presence of silica sand, silica gel, volcanic lava, or alumina. The reacted gas was then dissolved in aqueous ammonia and the solution was heated in a sealed bottle at 75 C. The ammonia and water were removed by evaporation; the residue was hydrolyzed with hydrochloric acid and passed through an amino acid analyzer. Harada and Fox (*ibid.*) reported that glycine, alanine, glutamic acid, aspartic acid, leucine, valine, proline, serine, isoleucine, threonine, tyrosine, phenylalanine, and alloisoleucine were found in the product. The formation of the basic amino acids was not fully studied, but peaks corresponding to lysine (or ornithine) and arginine were present; the possible presence of histidine was obscured by ammonia. Since sulfur was not present in the experimental mixtures, any possibility for the production of cystine or methionine was excluded.

Oró (1965) has reviewed experiments along similar lines which led to the detection of aspartic acid, asparagine, threonine, serine, glutamic acid, glycine, alanine, isoleucine, leucine, tyrosine, phenylalanine, lysine, and arginine in the reaction products. Formaldehyde, cyanide, hydroxylamine, hydrazine, and ethane were among the raw materials used, in addition to ammonia, methane, and water. The production of purines and pyrimidines by the action of heat on ammonium cyanide has been reported (*ibid.*; Oró, 1963). Ponnamperuma (1965) found spots corresponding to adenosine, adenylic acid, and adenosine di- and triphosphates in chromatograms of simpler reactants that had been treated with ultraviolet light. The ATP fraction was active in enzymatic identification tests. Ponnamperuma and Mack (1965) also reported the presence of guanylic, uridylic, and cytidylic acids.

Let us now, in a few inadequate sentences, review the conjectures as to what happened to the nutrients in the primeval nutritive soup after the great event occurred in which life made its appearance. The usual model is that so-called "living" organisms were formed by some sort of condensation of the nutrients into aggregations; presumably, these organisms were based on nucleic acid–protein systems. As the food material in the oceans became used up, enzymatic systems appeared for resynthesizing the nutrients from the products of catabolism, using either light or chemical energy derived from the breakdown of their substances to drive the synthetic processes forward. Essential to any such model is the evolution of informational molecules, which carry the coded information for the synthesis of biologically active proteins—the enzymes. The enzymes must have the necessary amino acids arranged in specific sequences.

The earliest forms of life, according to this model, were heterotrophic and obtained their supply of purines, pyrimidines, and amino acids from the primitive soup. It seems likely that the complex triplet coding system that serves for protein synthesis today must have evolved from simpler beginnings in these primitive organisms, rather than having been used from the outset. It is attractive to speculate that a step preceding the triplet code was a doublet code. The thread of doublets running through the existing triplet code lends substance to such a presumption.

It is of further significance to note that certain amino acids are derived from other amino acids by rather elaborate processes which indicate an evolutionary history:

Methionine is formed in heterotrophic organisms as shown in

Fig. 2-9. The methylation of homocysteine requires the presence of folic acid and, in some organisms, vitamin B_{12}, two relatively complex molecules. The production of methionine may, therefore, be regarded

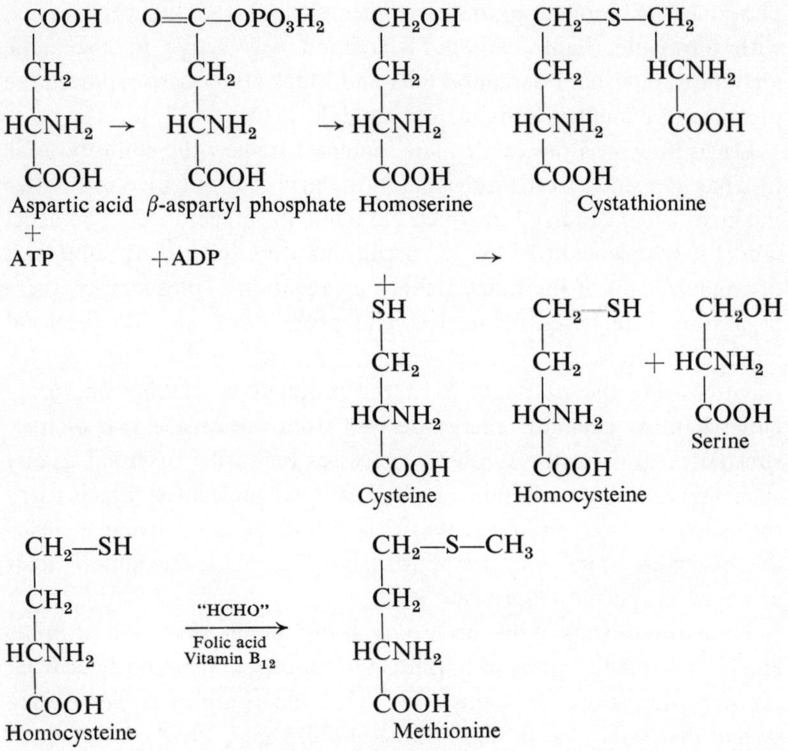

$$\begin{array}{cccc}
\text{COOH} & \text{O}=\text{C}-\text{OPO}_3\text{H}_2 & \text{CH}_2\text{OH} & \text{CH}_2-\text{S}-\text{CH}_2 \\
| & | & | & | \qquad | \\
\text{CH}_2 & \text{CH}_2 & \text{CH}_2 & \text{CH}_2 \quad \text{HCNH}_2 \\
| & | & | & | \qquad | \\
\text{HCNH}_2 \rightarrow & \text{HCNH}_2 & \rightarrow \text{HCNH}_2 & \text{HCNH}_2 \quad \text{COOH} \\
| & | & | & | \\
\text{COOH} & \text{COOH} & \text{COOH} & \text{COOH}
\end{array}$$

Aspartic acid β-aspartyl phosphate Homoserine   Cystathionine
+
ATP   +ADP   →

$$\begin{array}{ccc}
\overset{+}{\text{SH}} & \text{CH}_2-\text{SH} & \text{CH}_2\text{OH} \\
| & | & \\
\text{CH}_2 & \text{CH}_2 & +\text{HCNH}_2 \\
| & | & | \\
\text{HCNH}_2 & \text{HCNH}_2 & \text{COOH} \\
| & | & \text{Serine} \\
\text{COOH} & \text{COOH} & \\
\text{Cysteine} & \text{Homocysteine} &
\end{array}$$

$$\begin{array}{ccc}
\text{CH}_2-\text{SH} & & \text{CH}_2-\text{S}-\text{CH}_3 \\
| & & | \\
\text{CH}_2 & \xrightarrow{\text{``HCHO''}}_{\substack{\text{Folic acid} \\ \text{Vitamin B}_{12}}} & \text{CH}_2 \\
| & & | \\
\text{HCNH}_2 & & \text{HCNH}_2 \\
| & & | \\
\text{COOH} & & \text{COOH} \\
\text{Homocysteine} & & \text{Methionine}
\end{array}$$

Fig. 2-9

as a sophisticated biological feat in which two other amino acids, cysteine and aspartic acid, participate.

Tryptophan is the rarest of the 20 amino acids concerned in protein synthesis and the proteins of bacteria are particularly deficient in it. It is also the least stable of the amino acids to acid or alkaline conditions. Its biosynthetic origin involves the participation of serine with indole or indole-3-glycerol phosphate as shown in Fig. 2-10.

Glutamine is formed from glutamic acid, ATP, and ammonia by the glutamine synthetase reaction.

Asparagine is formed by ammonia and aspartic acid in a similar reaction or by aspartic acid and glutamine in a transamination reaction.

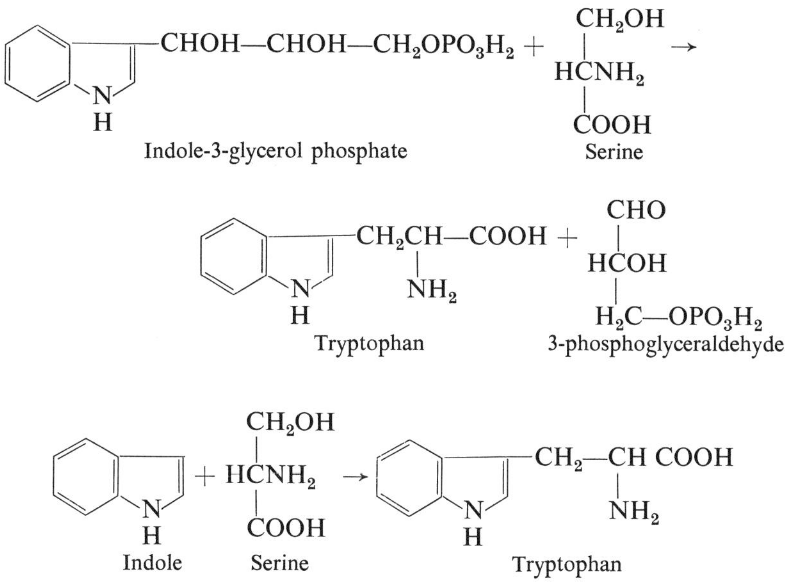

$$\text{Indole-3-glycerol phosphate} - CHOH - CHOH - CH_2OPO_3H_2 + \text{Serine}(CH_2OH - HCNH_2 - COOH) \rightarrow$$

$$\text{Tryptophan} (-CH_2CH(NH_2)-COOH) + \text{3-phosphoglyceraldehyde} (CHO - HCOH - H_2C-OPO_3H_2)$$

$$\text{Indole} + \text{Serine}(HCNH_2, CH_2OH, COOH) \rightarrow \text{Tryptophan}(-CH_2-CH(NH_2)COOH)$$

Fig. 2-10

Tyrosine is produced by the hydroxylation of phenylalanine in the phenylalanine hydroxylase reaction. A hydroxylated pteridine participates in this reaction. Tyrosine has also been reported to be produced abiogenically (Harada and Fox, 1965, p. 187).

Following these various considerations, we have speculated that the underlying pattern in the genetic code represents a vestigial survival of a primitive doublet code (Jukes, 1965a). This pattern appears when the 16 possible two-letter permutations of A, C, G, and U doublets are arranged in order and are compared with some of the triplets whose function has been determined or inferred (Table 2-15).

Five of the 20 amino acids that participate in protein synthesis are formed biologically from five of the remaining 15, as described above.

68 The Genetic Code

Of these remaining 15, all but two have been found to occur in the products obtained by heating mixtures of simpler chemicals. One of the two is histidine, whose possible presence in such mixtures is difficult

Table 2-15. A Proposed Archetypal Genetic Code and Changes That Relate It to Present Assignments

Quartets	Archetypal Assignments	Changes (If Any)	Present Assignments
AAd	lys	AAb	asN
		AAe	lys
ACd	thr	...	thr
AGd	ser or arg	AGb	ser
		AGe	arg
AUd	ilu	AUb	ilu
		AUA	ilu
		AUG	met
CAd	his	CAb	his
		CAe	glN
CCd	pro	...	pro
CGd	arg	...	arg
CUd	leu	...	leu
GAd	asp or glu	GAb	asp
		GAe	glu
GCd	ala	...	ala
GGd	gly	...	gly
GUd	val	...	val
UAd	End	UAb	tyr
		UAe	End
UCd	ser	...	ser
UGd	cys	UGb	cys
		UGG	try
UUd	phe	UUb	phe
		UUe	leu

b = C, U; d = A, G, C, U; e = A, G

to detect, as the methods for testing for histidine also respond to ammonia. The other is cysteine, which contains sulfur. Sulfur has not yet been used in the reaction mixtures, so the formation of cysteine is an undecided question.

The crude tarry mixtures formed by heating methane, ammonia, water, cyanide, etc., together are materials that would be considered unattractive by organic chemists, who traditionally prefer to work with pure substances. The presence of traces of biologically interesting compounds in such mixtures has come to light through the newer methods of analysis, especially chromatography in its various modifications.

The archetypal doublet code proposed elsewhere (Jukes, 1965a) is more succinctly outlined as consisting of 16 quartets of triplets AAd, ACd, etc., in which the third base is A, C, G, or U used synonymously and interchangeably (Table 2-15). The 16 assignments for the quartets are in the second column of the table, and the present code is shown in the third and fourth columns. It is also presumed that there could have been only 16 sRNAs—one for each quartet. Eight quartets have retained their original assignments; seven of the others have each kept either a purine-terminated or pyrimidine-terminated pair of triplets for the original amino acid and have lost the other pair to another amino acid during evolution. In this way, two codes for each of four "new" amino acids have been provided. The sixteenth quartet appears to consist of three codes for ilu and one for met, which is perhaps the newest of the amino acids that participate in the synthesis of polypeptides upon ribosomes. Methionine is in higher organisms the parent substance of the methyl groups of choline, which is absent from or present only in traces in many microorganisms (Goldfine and Ellis, 1964). Other microorganisms that contain choline do not synthesize it from methionine (*ibid.*). Methylation by methionine is also responsible for other biochemical products that may be regarded as comparatively sophisticated. Examples of these are creatine and epinephrine and the bases that are methylated after DNA and RNA have been synthesized; significantly, methionine does not provide the methyl group of thymidine, a more primitive substance, which, like methionine, acquires its methyl group from formaldehyde or formate via folic acid.

No attempt has been made to guess whether it was serine or arginine that was originally coded by the AGd quartet, or why serine, arginine, and leucine now have six codes. Perhaps aspartic and glutamic acids, owing to their similarity, both had the same sRNA and code and were used interchangeably in the primitive proteins.

Changes in the genetic code should presumably come about by changes in sRNA. Let us suppose that CAA was at one time one of the codes for histidine and that the corresponding sRNA, containing a complementary coding triplet UUG, underwent a mutational change in the region of its amino acid recognition site, so that it became charged with glutamine rather than histidine, glutamine being an amino acid that previously had no recognition site in any sRNA. The result would be the introduction of glutamine in proteins replacing histidine at all messenger-RNA sites containing the triplet CAA. Other sites, such as CAC and CAU, would retain histidine. If an organism could survive this change, it would have acquired a "new" amino acid, which might be an evolutionary advantage. We must assume that the organism would be quite primitive and simple to endure such a disruption. It is of interest that Braunitzer (1965) reported the occurrence of proflavine and 2,7-diaminofluorene mutants of bacteriophage *fd* which survived the introduction of a new amino acid, histidine, into their coat protein. The coat protein of the parent strain contained no histidine.

It is presumed that the evolutionary step that led to the adoption of the present code was *preceded* by the development of enzyme systems that could produce new amino acids. These amino acids were not present in the enzymes that produced them. This concept is illustrated by the fact that tryptophan synthetase of *E. coli* contains no tryptophan (Henning et al., 1962). New amino acids may, however, be added to proteins by single-base changes occurring as mutations, for example, AUA/AUG, ilu to met.

What is the meaning of the amino acid code? What is meant by the statement that GCC is a code for alanine?

The code places the amino acid adaptor molecules (the sRNAs) in the correct positions during protein synthesis. There is no known stereochemical or complementary affinity between alanine and the sequence –guanyl–cytidyl–cytidyl– in an RNA molecule. The relationship of GCC to alanine is due to the existence of an adaptor molecule which will connect alanine with GCC. This molecule is *alanyl transfer RNA*. As present in yeast, it has been shown to contain a known and ordered sequence of 77 nucleotides (Holley et al., 1965). Within this sequence there are two distinctive regions that are essential to the identification of alanine and to placing it in the genetically ordered loci in polypeptide chains. The first of these regions is an unidentified group of nucleotides which recognizes the complex consisting of alanyl adenylate bound to alanyl-activating enzyme, so that the enzyme transfers the alanine to ester linkage with the 3'-OH group of the terminal adenine group of uncharged alanyl-sRNA, thus charging the molecule. The second region consists of the sequence IpGpC at sites 36, 37, and 38, which complements with a triplet in mRNA.

Sequences of IGC, IAC, and IGA have been found in the sRNAs for ala, val, and ser, respectively (Holley et al., 1965; Armstrong et al., 1964; Dütting et al., 1965). These would correspond to the codes GCU, GCC, and GCA for ala; GUU, GUC, and GUA for val; and UCU, UCC, and UCA for ser. A second sRNA with a sequence CXY instead of IXY would carry the amino acid to the messenger triplet terminating with G in each case, while a third would have the sequence UXY to pair with the messenger triplets terminating with A and G, and a fourth would have the sequence GXY, pairing with the messenger triplets ending with U and C. In the case of coding triplets of the type in which XYU and XYC code for one amino acid while XYA and XYG code for a different amino acid, it is necessary to postulate that the pairing procedure can distinguish between XYA as compared with XYU and XYC. In these cases, an sRNA coding triplet could start with G to complement with either U or C in the third position of the messenger triplet, while the anticodon of a second sRNA could start with A to complement with U. The third sRNA would have an anticodon starting with U to complement with A and G, and the fourth would have C in the first position of the anticodon, to pair with G. The postulations for ambiguity are based on the "wobble" hypothesis (Crick, 1965), assuming that the third base in the messenger coding triplet may exhibit ambiguous pairing as follows: either U, C, or A will pair with I in position 36 of sRNA; either U or C with G; either A or G with U; U with A; and C with G. As an example, let us consider the case of aspartic and glutamic acids. One aspartyl sRNA should contain a coding triplet GUC to complement with GAU and GAC, and a

72   The Genetic Code

Table 2-16. Postulated List of sRNA (Transfer RNA) Coding Triplets to Complement with Messenger RNA Coding Triplets (See Table 2-13)

Amino Acid	Triplets Messenger RNA	Triplets Transfer RNAs	Amino Acid	Triplets Messenger RNA	Triplets Transfer RNAs
phe	UUU, UUC UUU	GAA AAA	Ilu	AUU, AUC AUU, AUC, AUA	GAU IAU
leu	UUA, UUG UUG	UAA CAA	met	AUG	CAU
ser	UCU, UCC UCU, UCC, UCA UCG UCA, UCG	GGA IGA CGA UGA	thr	ACU, ACC ACU, ACC, ACA ACG ACA, ACG	GGU IGU CGU UGU
tyr	UAU, UAC UAU	GUA AUA	asN	AAU, AAC AAU	GUU AUU
Gap	UAA, UAG UAG	UUA CUA	lys	AAA, AAG AAG	UUU CUU
cys	UGU, UGC UGU	GCA ACA	ser	AGU, AGC AGU	GCU ACU
try	UGA (?) UGG	UCA (?) CCA	arg	AGA, AGG AGG	UCU CCU
leu	CUU, CUC CUU, CUC, CUA CUG CUA, CUG	GAG IAG CAG UAG	val	GUU, GUC GUU, GUC, GUA GUG GUA, GUG	GAC IAC CAC UAC
pro	CCU, CCC CCU, CCC, CCA CCG CCA, CCG	GGG IGG CGG UGG	ala	GCU, GCC GCU, GCC, GCA GCG GCA, GCG	GGC IGC CGC UGC
his	CAU, CAC CAU	GUG AUG	asp	GAU, GAC GAU	GUC AUC
gIN	CAA, CAG CAG	UUG CUG	glu	GAA, GAG GAG	UUC CUC
arg	CGU, CGC CGU, CGC, CGA CGG CGA, CGG	GCG ICG CCG UCG	gly	GGU, GGC GGU, GGC, GGA GGG GGA, GGG	GCC ICC CCC UCC

second should contain AUC to complement with GAU. There should be two glutamyl sRNAs, one containing UUC to complement with GAA and GAG and the other containing CUC to complement with GAG. Sixty-three sRNAs would be required in terms of this scheme. It is assumed that the base corresponding to position 36 of Holley's yeast alanyl sRNA (Fig. 2-4a) will be A when the molecule is first formed by the action of RNA polymerase. A specific *anticodon deaminase* is postulated to change this A to I in the case of this sRNA and of the analogous sRNAs for ser, leu, pro, arg, ilu, thr, val, and gly. The scheme is set forth in Table 2-16.

The two triplets, UAA and UAG, which for brevity are designated as coding for gaps, are evidently the signals for polypeptide chain termination (Brenner et al., 1965; Weigert and Garen, 1965). The mechanism for release must include hydrolysis of the ester linkage between the 3'-OH of adenosine and the α-carboxyl group of an amino acid.

It seems probable that the genetic code is universal and constant, within the biological limits of these categorical terms. There are several lines of evidence for this conclusion. Synthetic polyribonucleotides incorporate the same amino acids in cell-free systems prepared from various organisms. Many viruses can each infect more than one host species; thus, it is implied that a common protein-synthesizing mechanism in several hosts is at the disposal of the parasite when it enters the cell as a single DNA or RNA molecule. The amino acid sequence of 11 consecutive residues which is common to all the cytochromes *c* (Chapter 6) that have been examined could scarcely have persisted for several hundred million years unless the code remained unchanged and identical in all the species involved.

How could changes in the code take place? This can be examined by taking a hypothetical example: Could the AAG code for lysine change to ACG? Obviously this change could not take place in the structural genes or in messenger RNA, since it would not be possible for all the AAG sequences to change simultaneously to ACG. If one such sequence changed, the effect would be the substitution of threonine for lysine at a single locus in a single protein, a mutation that is known in one of the hemoglobin variants and in TMV.

The other possibility is that the presumed coding sequence CUU in the lysyl-sRNA that corresponds to AAG would become changed by a mutation to CGU. This change would result in lysine being substituted for threonine at loci corresponding to ACG messenger codes throughout the proteins of the mutant. The new lysyl-sRNA would presumably compete with threonyl-sRNA for such sites, while the AAG triplets would be translated by the other lysyl-sRNA. Perhaps very primitive and simple organisms could survive such a sweeping change, but in terms of the complex forms of life that exist at present, one would expect that this would have lethal consequences, although it is true that suppressor strains of *E. coli* can insert serine, glutamine, or tyrosine at some of the chain-terminating loci in the amber mutants (Brenner et al., 1965; Weigert et al., 1965). (See Chapter 4.) It would seem that the chemical processes of heredity tend to maintain constancy and universality in the terrestrial genetic code. We must search in other environments if we expect to find different codes.

[3]

Microbiology and the Study of Heredity

... a nucleic acid of the desoxyribose type is the fundamental unit of the transforming principle of Pneumococcus type III.

O. T. Avery, C. M. Macleod, and M. McCarty,
J. Exptl. Med. 79:137 (1944).

It is difficult to separate subject matter dealing with protein synthesis and evolution into categories, and the topics in each chapter tend to recur elsewhere in this book. In this chapter, some of the recent applications of microbiology to the problems of biochemical genetics will be discussed, including sRNA, DNA replication, RNA and ribosomes, strand selection, and gene clusters.

The inexorable progress of biochemistry, even as it catalogs the differences between the varied forms of life, continually reveals their underlying similarities. The contrast between higher animals and plants, obvious at a glance, becomes satisfyingly documented for us by the descriptive achievements of science; differences in structure, composition, and function between the two kingdoms extend into every cell. Within each cell the mechanisms for this differentiation are indelibly and invisibly encoded in a sequence of bases and are translated perceptibly into the phenotype by the protein-synthesizing apparatus. Microorganisms, originally described as the "little animals" of Leeuwenhoek, the yeasts, and especially the bacteria, contain this machinery and have been extensively used as source material in studies of DNA and its replication, of the various forms of RNA, and of cell-free protein synthesis. From these studies conclusions have been drawn as to the general nature of the genetic message and its translation. Among the advantages of bacteria for these studies are the availability of techniques for fractionating the contents of their cells, the ease with which they can be rapidly produced in large numbers, and the absence of many complexities of higher organisms, such as alternation of generations, separate nuclear structure, mitochondria, plastids, and differentiation.

The many differences between bacteria and multicellular organisms in intracellular structure and organization are amply discussed in various textbooks and reviews (e.g., Brachet and Mirsky, eds., 1958–64; Murray, 1962; Stanier, 1964; Stanier and van Niel, 1962). The important similarities include the use of the DNA-RNA-ribosome system in the storage and expression of genetic information. The genetic code appears to be the same in all organisms.

Studies with bacteriophages have ramified so extensively as to have become a separate branch of biochemical genetics. The field is well reviewed by Luria (1962) and Stent (1963). Bacteriophages have numerous advantages for the study of biochemical genetics, including their rapid multiplication, the presence of enormous numbers of them in a small volume of culture medium, the speed with which aberrant or mutant types may be found by plating out on agar seeded with bacteria, their susceptibility to mutagens, the phenomena of transduction, lysogeny, recombination, host-range specificity, the "early" and "late" protein phenomena, temperature sensitivity, chromosomal map circularity, and gene mapping.

Bacteriophages are better subjects for the exploration of DNA than of proteins, since the proteins synthesized by phages are present only in minute amounts in contrast to such proteins as the hemoglobins, cytochromes, and TMV protein, which may be prepared pure in comparatively large quantities for the studies of their amino acid sequences. The studies with bacteriophages have suffered from the lack of knowledge of the translation of the genetic phenomena that were observed. It was not until Sarabhai et al. (1964) collected sufficient quantities of phage head protein from the T4D amber mutants for peptide studies that any direct relationships were found between mutations in bacteriophages and sequence of polypeptides in phage proteins. Subsequent publications by Brenner, Stretton, and Kaplan (1965) reported the use of phages to identify the interval triplets, or gaps, UAA and UAG, which signal when a protein molecule should be terminated by freeing the carboxyl group of the terminal amino acid from its combination with sRNA, and by Terzaghi and coworkers (1965) to relate the deletions of a base to a change in the "reading frame."

Transfer RNA (sRNA) and Transfer Enzymes

This family of molecules was discussed in Chapter 2. Crude preparations of mixed sRNAs are readily obtained from various tissues and organisms, including liver, heart, tumors, tissue cultures, *E. coli*, and yeast. The last-named has been the principal source for large-scale fractionation studies.

To recapitulate, every organism contains a family of sRNA molecules, each of which may be represented by the symbol PQR to designate its three separate functional regions without reference to their relative juxtapositions, P referring to the sequence of three bases that pairs with a complementary coding triplet, Q a recognition site which accepts a specific amino acid brought to it by an activating enzyme (aminoacyl-sRNA synthetase) for attachment to the terminal A-3'-OH, and R the remainder of the molecule serving other functions, such as that of combining with ribosomes. If PQR represents a yeast alanyl-sRNA, the current theory is that there will be a corresponding *E. coli* alanyl sRNA containing the same coding triplet P, such as IGC, but a somewhat modified Q region that may or may not be recognized by the yeast alanine-activating enzyme, depending on evolutionary changes that may have caused a divergence. The third region, R, may also have undergone evolutionary differentiation but seems to be conserved at least in part, for the findings by Zamir, Holley, and Marquisee (1965) indicate that there is a common sequence $GpTp\psi pCpGp$ in all the sRNAs so far examined. The possibility is attractive that this sequence is concerned with some function that is shared by all sRNAs, such as ribosomal binding.

The attachment of an amino acid to an sRNA of one organism by the activating enzymes of a different organism has been of much interest to various investigators, and "cross-reactions" have been described between the sRNAs of one organism and the activating enzymes of another (Allen et al., 1960; Berg et al., 1961; Hecht, Stephenson, and Zamecnik, 1959; Rendi and Ochoa, 1961). The subject has been explored with great assiduity by Sueoka and his collaborators, and their investigations will be reviewed as a good illustration of the use of microbiology in studies of evolutionary differences between species.

Essential to such studies is the separation of different sRNAs for the same amino acid; for example, if an activating enzyme from yeast attaches leucine to *E. coli* sRNA, it is first necessary to separate the *E. coli* leucyl sRNAs to indicate which are being charged with leucine by the heterologous enzyme. Fractionation with methylated albumin columns was used by Yamane and Sueoka (1963) for this separation. They encountered three situations commonly in cross-reactions with a heterologous enzyme: (1) no charging; (2) it charged just as well as the homologous enzyme; (3) it did not charge all the components. A fourth possible situation was encountered in one instance: *E. coli* synthetase charged a yeast sRNA with leucine, but the charged sRNA had different properties from any of the regular yeast leucyl-sRNAs and amounted to only 1 percent of the normal total fraction.

Another example of the fourth situation was reported by Barnett and Jacobson (1964). They found that the phenylalanine-activating enzyme from *Neurospora crassa* attached phenylalanine to the following *E. coli* sRNAs: (1) one that was chromatographically indistinguishable from normal *E. coli* phenylalanyl sRNA; (2) one to which the enzyme could attach either phenylalanine or alanine; and (3) one to which the *Neurospora* enzyme but not the *E. coli* enzyme could attach phenylalanine. These results are best explained on the basis that the activating enzymes and the recognition sites *both* differ among species. Such a situation would be in accordance with other evolutionary changes in the sense that it could be brought about by a series of changes in the base sequences of the genes that are responsible for sRNA and for activating enzymes. The changes would be parallel in the sense that a base change in the recognition site of a specific sRNA would have to be followed by a compensatory change in its activating enzyme. Some degree of flexibility in the functioning of the enzyme would be needed during the intervening period. The problem is compounded if a change takes place in the recognition site of only one member of a family of sRNAs that carry a single amino acid and are charged by a single activating enzyme. This appears to be the case in *E. coli*; only one enzyme is responsible for the charging of all *E. coli* leucyl-sRNAs (Keller and Anthony, 1963; Sueoka, 1965; Yamane and Sueoka, 1964). Similar findings were reported by Yamane and Sueoka for a number of other synthetases and sRNAs; in every case a single enzyme appeared to be responsible for charging all the sRNAs for one amino acid. Furthermore, in the presence of leucyl-sRNA synthetase, leucine can shift from one sRNA to another, thus emphasizing the functional identity of the recognition sites in all the *E. coli* leucyl-sRNAs.

The information that enables development and differentiation of living organisms to take place is carried in DNA. The procession of the embryo through its protean recapitulation of biological history; the measured and finely adjusted formation of organs, systems, and appendages; the flowering of sexual maturity; the production of sperm and ova; the onset of ripening and senescence; the rhythms of life—all these are regulated and ordered by inherent chemical messages whose complex nature is at present hidden.

The smallest and simplest known living forms that go through a life cycle are the bacteriophages. They inject their DNA in a bacterial cell, and in a matter of minutes the functions of the host are usurped and new bacteriophage particles are synthesized, assembled, and liberated by lysis. The procedures concerned with this rapid cycle have been charted on the circular genetic map of bacteriophage T4 (Edgar, 1962), which lists 48 consecutive genes, or *cistrons*, each specific for the coding of a single protein. The early enzymes are produced at 37 C during the period that starts 3 minutes after infection and ends 7 to 9 minutes later. The synthesis of T-even phage DNA, characterized by its content of hydroxymethyl cytosine replacing cytosine, is carried out by a "late" protein, and, together with the synthesis of structural proteins, starts at 7 to 9 minutes and continues until maturity at 25 minutes. In T2-infected cells, more than 60 percent of the late protein synthesized after 12 minutes is phage head membrane protein. Luria (1962) has pointed out that the early and late functions of T4 represent two mutually exclusive sets of activities, a finding that may hold the key to problems of the coordinated regulation of gene action in processes such as differentiation.

The change in the pattern of protein synthesis following infection with bacteriophage T2 was investigated by fractionation of *E. coli* sRNAs of various time intervals (Sueoka and Kano-Sueoka, 1964). The patterns of elution of 17 different sRNAs from methylated albumin columns were examined. This procedure revealed changes that occur following phage infection in the relative amounts of the sRNAs for the same amino acid. In only one case, that of leucine, was such a change found; the front component (leu I) in the elution of leucyl sRNA decreased to one-half the noninfected value (*ibid.*). Leu I has three components. The leucyl-sRNA synthetase was not changed.

The rate of change in leucyl-sRNA was measured at 3, 5, and 8 minutes following infection. A new component appeared at 3 and 5 minutes and disappeared at 8 minutes. The new component could be charged with yeast leucyl-sRNA-activating enzyme, which charges leu I.

The alteration in leu I produced by T2 bacteriophage was abolished by chloramphenicol, indicating that the prevention of protein synthesis stopped the change from taking place.

Sueoka proposed that the time-related appearance and disappearance of the new leucyl-sRNA peak is involved in the shift from early to late phases of protein synthesis following phage infection. This could come about if certain messenger sites for leucine—let us say, CUA sites—are translated in the formation of phage proteins during the 3-to-5-minute period following infection. The necessary specific leucyl-sRNA corresponding to the CUA code then disappears or is modified. We do not know how or why; it was not found at 8 minutes in Sueoka's experiments. The result is that CUA sites are no longer translated, and the corresponding proteins, the early proteins, are not synthesized, thus shutting off the early gene. Sueoka proposes this *adaptor modification* hypothesis as a major principle in differentiation.

Microbiology in the Study of Replication of DNA

Most of our understanding of the replication of DNA on an experimental basis is derived from microbiological studies. *E. coli* and its bacteriophages have been prominent in these investigations.

One of the first landmarks in such research was the report by Meselson and Stahl (1958), who grew *E. coli* in a medium rich in compounds containing the heavy isotope N^{15}. This procedure led to

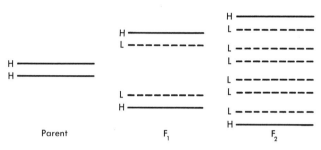

Fig. 3-1. Replication of heavy DNA in light medium.

the incorporation of N^{15} in the DNA, which increased its density so that it could be shown to take up a characteristic position when suspended in cesium chloride in the cell of the analytical ultracentrifuge, differing from the position reached by unlabeled DNA. The bacteria were then transferred to an N^{14} (light) medium and allowed to replicate for one generation. The DNA was removed from the cells, and its density was measured by ultracentrifugation as described. All the sample had a density corresponding to that of a hybrid strand, one half containing N^{15} and one N^{14}. After another replication, two bands were found in the photograph of the density gradient, indicating an equal division between "heavy-light" and "light-light" DNA molecules. The findings fully substantiated the Watson-Crick model of DNA replication and may be represented as in Fig. 3-1. Note that heavy-heavy strands are never re-formed; after the first replication the hybrid strands persist and are diluted at each cell division by light-light strands. From the evolutionary standpoint, it can be seen that the parental (heavy) strands persist indefinitely, although they are diluted after n generations according to the formula $1:2^n - 1$.

Confirmation of this model on a genetic basis was obtained by Yoshikawa and Sueoka (1963a), who used transformation in *B. subtilis*, and actual visualization of the process was shown in *E. coli* by Cairns (1963).

The order of replication of genes within the chromosome was demonstrated and mapped by the use of transformation, the uptake by recipient cells of genes carried in a small piece of DNA obtained from a donor cell. This procedure was used by Yoshikawa and Sueoka (1963a) in their experiments on chromosome duplication. The normal, or "wild," type of *B. subtilis* is separated from a rapidly growing culture, and the DNA is extracted from the cells by gentle lysis followed by purification. The long DNA molecules obtained by this procedure are broken into pieces of moderate length by shearing. The preparation is then incubated with another strain of *B. subtilis*. This strain is a defective mutant which has lost the ability to synthesize three or more essential metabolites such as adenine, methionine, and leucine or isoleucine, methionine, and leucine. The genes, or gene clusters, that code for the enzymes synthesizing these essential metabolites are mapped as follows:

1. The recipient cells are shaken with donor DNA for 40 minutes. Following this the DNA remaining in the solution outside the cells is destroyed by addition of deoxyribonuclease. During this period some of the donor DNA enters the recipient cells, carrying with it the genes that will repair their deficiencies. These genes now become incorporated into the DNA of the recipient cells, which are allowed to grow on a culture medium that is deficient in one of the essential metabolites. The number of bacteria that have been transformed to self-sufficiency is detected by this means.

2. Twice as many bacteria were transformed to adenine-synthesizing ability as were transformed to methionine-synthesizing ability if the DNA was obtained from rapidly growing wild-type *B. subtilis*. In contrast, when the DNA was obtained from resting wild-type cells, the "adenine-methionine ratio" was 1:1.

By means of these and similar experiments, it was concluded that the total DNA, or chromosome, of *B. subtilis* was a long strand that replicated from the same starting point preceding each cell division. The "adenine" gene was near the starting point and the "methionine" gene was near the terminus, as shown in Fig. 3-2.

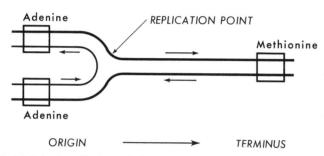

Fig. 3-2. Model of replication of the chromosome of *B. subtilis*, showing the presence of two genes for synthesis of adenine and one for synthesis of methionine when the gene is partially replicated (from Yoshikawa and Sueoka, 1963a).

It had previously been concluded that the chromosome of *E. coli* is a circular molecule of DNA (Jacob and Wollman, 1957) on the basis of mapping the linkage of its genetic markers. The replication of a circular piece of DNA, in terms of the model shown in Fig. 3-2, should appear as in Fig. 3-3. This figure is actually drawn from a photomicrograph published by Cairns (1963), who grew *E. coli* on a medium containing radioactive (tritiated) thymidine. The DNA was removed from the cells by gentle lysis and placed on photographic plates, where it was left for several weeks. The illustration in Fig. 3-3 shows a circular DNA molecule in the act of replication; the region A is the partially completed new molecule. The picture indicates that the replicative act starts at a single locus that travels along the chromosome.

The new strands of DNA are copied first from the original strands and then from each other by complementary base pairing, of course, so that there is rapid dilution of the original material by successive cell divisions, each of which is preceded by a doubling of the DNA molecule. It is of great importance from the standpoint of evolution to know how accurately the copying process is carried out and how completely the information is preserved in the time intervals between cell divisions.

The copying procedure and the preservation of the DNA base sequence are both subject to evolutionary pressures. Nature, like a proofreader, eliminates "bad copy." Changes in the DNA that produce defective offspring are, for this very reason, not perpetuated. There is also, however, a mechanism that actually repairs the strands of DNA when they are damaged by certain disruptive processes. One such process is the action of ultraviolet light. This produces "thymine dimers" in adjoining T residues, which block DNA synthesis in vivo and in vitro. They have the structural formulas typified by

$$\begin{array}{c}\text{structural formula of thymine dimer}\end{array}$$

Enzymes exist both for removing thymine dimers from the strands of DNA and replacing them with unchanged thymine molecules, thus restoring the original structure and coding function of the strand.

This process, which takes place either in the light or in the dark, is termed *photoreactivation, photoprotection,* or *ultraviolet reactivation* and occurs in ultraviolet resistant *E. coli* cells (Setlow and Carrier, 1964; Boyce and Howard-Flanders, 1964). The genetic locus for the reactivation system has been found in the *Hfr* (high frequency of recombination) chromosome in *Hfr* strains of *E. coli.*

The repair mechanism is carried out by removing a number of adjoining bases in addition to the thymine dimer. The excised sequence is then replaced by the action of a DNA polymerase, which adds complementary bases as specified by the opposite strand, so that the molecule is restored to its previous condition.

Yoshikawa (1965) studied the synthesis of DNA during germination of *B. subtilis* spores. The addition of the antibiotic chloramphenicol, which stops protein synthesis, showed that a protein must be synthesized before each cycle of DNA replication in order to initiate the process of chromosome duplication. Chloramphenicol did not, however, stop a slow turnover of thymine in the DNA, which appeared to take place along the entire length of the chromosome even when growth was not

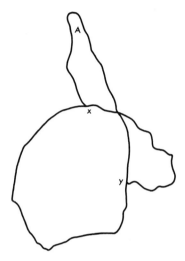

Fig. 3-3. Tracing of photomicrograph of replication in *E. coli* chromosome (from Cairns, 1963).

taking place. Apparently, a separate enzyme exists which brings this about. It may have the function of maintaining the integrity of the bacterial chromosome. Such a process could have a conserving action in preventing the "erosion" of DNA and the consequent loss or change of its genetic function. The effect is analogous to that of photoreactivation, although a direct comparison of the two phenomena has not been made.

The Interaction of Ribonucleic Acid with Ribosomes

The processes of heredity are carried out by means of translation of the genetic message into proteins on ribosomes. Much of the knowledge of the nature and behavior of ribosomes has come from studies with *E. coli* and yeast. The diameter of the typical *E. coli* ribosome is about 0.02 μ. The current theory of protein synthesis states that messenger RNA combines with ribosomes as a preliminary to the synthesis of proteins. The messenger strand is postulated to travel along the ribosomes (or vice versa) in a manner reminiscent of a motion picture film passing over sprockets of a projector. This concept was proposed in advance of experimental evidence to verify it. Studies with *E. coli* have provided some substance for the theory.

It was reported by Takanami and Okamoto (1963) that polyuridylic acid could be used as a "model messenger" in studies of the combination between RNA and ribosomes. A "heavy" complex was formed by attachment of several ribosomes to a single molecule of poly-U, as shown by the sedimentation pattern of the mixture in sucrose density gradients. In further experiments ribosomes were separated into their 50s and 30s components by dialysis against 0.25 mM Mg buffer. The 30s subunits readily combined with poly-U in complexes formed by the attachment of several ribosome particles to a single molecule of poly-U. No complexes were formed between the 50s subunits and poly-U, showing that the 30s subunit attaches to the messenger.

Several reports have described the formation of polysomelike structures in cell-free systems by synthetic ribonucleotides, including poly-U, poly-UC, and poly-UG (Barondes and Nirenberg, 1962; Spyrides and Lipmann, 1962; Gilbert, 1963). Takanami and Zubay (1964) were able to determine the length of the portion of RNA that is bound to the 30s unit of the ribosomes. They used the following procedure: Poly-U was incubated with ribosomes so that complexes were formed. Treatment of the complex with pancreatic ribonuclease digested most of the poly-U but left a fragment bound to each ribosome. This fragment had been protected against enzymatic digestion by interaction with the ribosome. The fragment of poly-U was eluted with sodium dodecyl sulfate and phenol, and the measurement of its average length showed a maximum corresponding to about 27 residues.

Poly-A, poly-C, poly-AU, and various forms of natural RNA, including viral RNA, did not combine readily with ribosomes (Okamoto and Takanami, 1963; Haselkorn and Fried, 1964). TMV RNA failed to produce the synthesis of TMV protein in cell-free systems obtained from *E. coli* (Aach et al., 1964). In contrast, the small RNA bacteriophage f2 has been shown to bring about amino acid incorporation in cell-free systems obtained from *E. coli* (Zinder, 1963). The f2 bacteriophage was used in studies of RNA binding of *E. coli* ribosomes (Takanami, Yan, and Jukes, 1965). The characteristics of the fragment of f2 RNA that was attached to ribosomes were examined by a procedure similar to that used by Takanami and Zubay (1964) in their experiments with poly-U. The nucleotide composition of the fragment was about 21 percent A, 25 percent C, 35 percent G, and 20 percent U as compared with 22.4 percent A, 26.5 percent C, 25.9 percent G, and 25.2 percent U for total f2 RNA. It was concluded that the ribosome was attached to a specific locus of the RNA. The fact that the composition of the fragment did not vary gave an opportunity for detecting whether motion of the ribosomes relative to messenger RNA took place during protein synthesis. This was examined by incubating the ribosome-f2-RNA complex with the necessary components of an amino-acid-incorporating system, following which the attached RNA fragments were isolated as before. The fragments obtained following incubation were found to have a nucleotide composition of about 26 percent A, 22 percent C, 29 percent G, and 23 percent U. The change in composition was thought to result from movement of the ribosome to another region of the RNA molecule because of the process of protein synthesis.

The Initiation of Polypeptide Chains

Waller and Harris (1961) noted that the proteins from ribosomes of *E. coli* contain certain characteristic NH_2-terminal amino acids. Methionine and alanine were about 85 percent of the total, and 13 percent of the remaining 15 percent were serine and threonine. This might have been due to a specific type of sequence characteristic of ribosomal proteins; however, it was found by Waller (1963) that the proteins of the supernatant fraction of *E. coli* extracted after ribosomes had been removed by centrifuging were similarly constituted. The same four NH_2-terminal amino acids predominated and together accounted for about 85 percent of the total. It could be inferred from these results that the codes for methionine and alanine were possibly "initiation codes" for protein synthesis in *E. coli*.

An impetus for further study of this problem was in the report by Marcker and Sanger (1964), who described the presence of N-formyl-methionine-sRNA in *E. coli*, formed by formylation after attachment of methionine to sRNA. Clark and Marcker (1965) found that this unusual sRNA was bound to *E. coli* ribosomes by either UUG or AUG in a trinucleotide binding system of the type described by Nirenberg and Leder (1964). Interest in N-formylmethionine was heightened by the fact that a formylaminoacyl-sRNA differs from an aminoacyl-sRNA in lacking the NH_3^+ group. This group is essential to polypeptide chain extension, but its absence might be necessary for binding to the ribosome at the initiating site and for aiding in the formation of the first peptide bond.

It was reported by two groups (Webster et al., 1966; Adams and Capecchi, 1966) that formylmethionine was the amino-terminal group of the coat proteins of f2 and R17 bacteriophages produced in cell-free systems. Webster et al. noted that the sequence of the coat protein differed in the cell-free system from the protein produced *in vivo*, which had an initial sequence of NH_2ala.ser.asN.phe.thr.g1N.phe.val. . . (Konigsberg et al., 1966). The conclusion was that the N-formylmethionine residue was removed by an enzyme *in vivo*. Adams and Capecchi reported that N-formylmethionine was followed by ala in the phage coat protein. They also found it present in the S20 fraction of R17 phage proteins.

The conclusions were that the initial signal for protein synthesis consisted either of a triplet coding for N-formylmethionine-sRNA or of an even longer sequence, perhaps nine bases, coding for N-formyl-met.ala.ser. In addition, there is an implication that the formylmethionine and perhaps the following alanine and serine residues are successively removed by an exopeptidase, their presence in diminishing quantities in the NH_2-terminal position reflecting the stepwise and progressively decreasing action of the enzyme.

At present there is no indication for extending this concept to organisms other than *E. coli* and its host-specific bacteriophages. Many proteins from other organisms start with amino acids other than met, ala, ser, or thr; and some proteins start with acetylamino acids (Chapter 6). The acetylation process was found to take place independently of, and presumably subsequent to, polypeptide synthesis (Marchis, Mouren, and Lipmann, 1965), and the enzyme transferred an acetyl group nonspecifically to the NH_2-terminal amino acid of several proteins.

Strand Selection of DNA by RNA Polymerase

Unlike DNA, RNA exists in single strands, even though these may be looped back on themselves into hairpinlike molecules as in the case of sRNA (Fig. 2-4). The existence of long single strands of RNA is

Fig. 3-5. Alternating transcription by RNA polymerase.

essential to the messenger concept, in which these strands are extended on ribosomes to receive the individual molecules of charged transfer RNA by complementary pairing during protein synthesis. The question remains of whether the messenger strands in living organisms are produced (1) from both strands of the entire length of the DNA molecule, (2) from one of the strands, and (3) alternately from staggered regions on each side as indicated in Fig. 3-5.

Possibility (3) is supported by the observation that some genetic regions are transcribed in opposite directions from others (Lewis, 1951). Possibility (1) makes for evolutionary complications. Let us suppose that a three–base coding sequence in one strand of DNA underwent a mutation from CCT to CTT. The messenger RNA triplet would change from AGG to AAG (reverse polarity), corresponding to arg to lys in the coded protein. In the opposite DNA strand, the change would be from AGG to AAG, and in the other messenger from CCU to CUU, pro to leu in a different protein. We assume that only a restricted number of amino acid changes in certain "variable" loci can be tolerated in a protein. This assumption follows from examples such as the cytochromes c (Chapter 6). A few such changes, which would be acceptable in one protein by evolutionary selection, would soon make the other protein biologically useless by the effect of their reverse counterparts on its amino acids. Evolution would tend to favor the survival of organisms that make messenger RNA by complementation of only one strand of DNA. Another argument against the transcription of both strands of DNA into RNA in vivo is that, except for viral RNA, which is replicated by a different enzyme, no-one has yet demonstrated the existence of RNA-RNA hybrids of natural origin, although it is known that RNA polymerase will transcribe both strands in vitro and thereby produce RNA molecules that can become double-stranded by RNA-RNA annealing (Geiduschek, Moohr, and Weiss, 1962). Possibility (2) remains, with its modification (3) as shown in the above diagram. The enzyme preferentially initiates RNA chains with a purine nucleotide (Maitra and Hurwitz, 1965; Krakow, 1965). This would favor single-strand selection of pyrimidine loci in the DNA template as starting points, since the other DNA strand would have a purine residue at these loci.

One indication of asymmetric transcription is that various samples of RNA have shown an inequality between G and C content. These include RNA produced in *E. coli* by bacteriophage infection (Volkin and Astrachan, 1956; Nomura, Hall, and Spiegelman, 1960), ribosomal RNA (Yankofsky and Spiegelman, 1962*b*), and RNA that hybridized with denatured DNA on columns (Bautz and Hall, 1962; Bolton and McCarthy, 1962).

These findings are not clear-cut, for various explanations might provide a basis for their interpretation, such as the absence of part of the total RNA from the material being examined.

The problem was approached by Hayashi, Hayashi, and Spiegelman

(1963) with the use of bacteriophage ϕX 174. This bacteriophage contains a circular molecule of single-stranded DNA which is composed of about 5,600 bases. The molecule has been photographed in the electron microscope (Freifelder, Kleinschmidt, and Sinsheimer, 1964). The existence of DNA in a single strand in this bacteriophage was at one time cited as being in conflict with the Watson-Crick model for replication of DNA, since the model calls for the simultaneous complementary pairing of the bases in two strands as an essential step in the transmission of genetic information. How, then, could a single strand duplicate itself by such a means to pass the information along to its offspring? Once again the apparent exception was shown to conform. ϕX 174 was found to produce a replicative form which is double-stranded in the *E. coli* host cells. The second strand of the replicative form is complementary to the infective strand by Watson-Crick pairing and produces a new infective single strand by a DNA-polymerase reaction. The infective form carries hereditary information but does so in a manner that is perhaps more economical of space and material than in the case of ordinary double-stranded DNA.

The replicative form also produces RNA by the RNA-polymerase reaction. In the experiments by Hayashi et al. (1963) the RNA generated during infection was separated from host RNA by pulse labeling and column chromatography so that it was possible to detect and fractionate a peak differing from that found in uninfected cells, reaching a maximum at 50 to 53 minutes post-infection. The RNA present in this peak was found not too complex with mature single-stranded ϕX 174 DNA but to hybridize with the heat-denatured replicative form according to the model

Infective strand	Replicative form	Replicative form
$\overline{\text{CATTG}}$	$\rightarrow \dfrac{\overline{\text{CATTG}}}{\text{GTAAC}}$	$\rightarrow \dfrac{\overline{\text{CATTG}}}{\text{GTAAC}}$
		CAUUG RNA

The base composition of the RNA thus corresponded to that of the infective form (U replacing T) and was complementary to that of the other DNA strand, as predicted from the above model.

Marmur and Greenspan (1963) reached similar conclusions, as follows: Bacteriophage SP8, which is virulent for *B. subtilis*, was found after denaturation to produce strands that were separable by centrifugation or chromatography. The heavy (H) strand is richer in pyrimidines, and the light (L) strand is richer in purines than the native bacteriophage DNA. RNA complementary to the H strand but not to the L strand was isolated from SP8-infected *B. subtilis*. The complementation was demonstrated by resistance of the RNA-HDNA hybrid strand to ribonuclease and by the trapping of RNA in agar

Table 3-1

	C	A	U (or T)	G
DNA, light strand	21.3	30.1	24.5	24.1
DNA, heavy strand	24.2	24.0	32.1	19.9
α-RNA	20.5	31	23	25.5

columns containing H-DNA. Tocchini-Valentini et al. (1963) used a mutant of bacteriophage α which grows in *B. megatherium*. They prepared α-RNA from the infected bacteria by annealing the total RNA with denatured DNA; only the α-RNA hybridized. The α-RNA was then separated from DNA by being melted and passed through a nitrocellulose filter, which retains single-stranded DNA. The α-RNA was then examined for its ability to anneal with either of the separate strands of α-DNA; these were separated into light and heavy strands by denaturation and chromatography on methylated albumin-kieselguhr columns. The α-RNA was found to anneal with the heavy strand and to resemble the light strand in base composition (Table 3-1).

As a further exploration, RNA complementary to both DNA strands (αC RNA) was synthesized in vitro by the RNA-polymerase reaction, with double-stranded αDNA as a primer. The αC RNA was mixed with αDNA, and the mixture was melted and allowed to reanneal. Hybrid RNA-DNA strands were formed.

Gene Clusters

There are a number of instances in which a group of enzymes is concerned with the successive steps in a metabolic pathway. The genes that supply the information for such a series of enzymes may often be closely linked on the map of the bacterial chromosome, as was shown by Demerec and Hartman (1959) for *E. coli* and *Salmonella*.

Demerec (1965) pointed out that these clusters do not occur in *Neurospora*, in which chromosome inversions occur frequently; and Horowitz (1965) noted that inversions of the DNA in bacteria, which possess a single chromosome rather than multiple chromosomes, would reverse the polarity of a segment of DNA, thus producing chaos by scrambling the base sequence. Demerec (1965) considers it probable that extensive gene clustering is confined to a certain group of bacteria, including the enteric group.

It was proposed by Jacob and Monod (1961) that a gene cluster may behave as a unit in the transfer of information under the control of an *operator gene*. The group containing the operator gene and the *structural genes* coding for the related enzymes is termed an *operon* (Jacob et al., 1960).

A cluster of nine genes supplying the information for the synthesis of histidine has been intensively studied by Ames, Hartman, and their coworkers. This cluster with its control mechanisms, the "histidine operon," has been termed the *histidon* (Ames and Hartman, 1963). In *Salmonella typhimurium* it consists of an operator gene ("O") and a continuous linear series of linked genes that produce a long strand of messenger RNA, the molecular weight of which, measured by ultracentrifugal methods, is equivalent to about 13,000 nucleotides (Martin, 1963). The messenger carries the coding information for 10 enzymes which catalyze the reactions necessary for the 11 steps in the biosynthesis of histidine shown in Fig. 3-4 (Loper et al., 1964). The genes are not arranged in the same sequence as the order of the chemical reactions. The biosynthesis starts with ATP and a purine precursor, phosphoribosyl pyrophosphate. A fragment of the pyrimidine portion of the purine ring of adenine is destined to be transformed into histidine by the series of chemical reactions catalyzed by these enzymes. To achieve this, the six-membered heterocyclic ring of adenine first attaches a ribose phosphate group at position 1 to form phosphoribosyl-ATP (PR-ATP). The ring then opens, and the imidazole portion of the

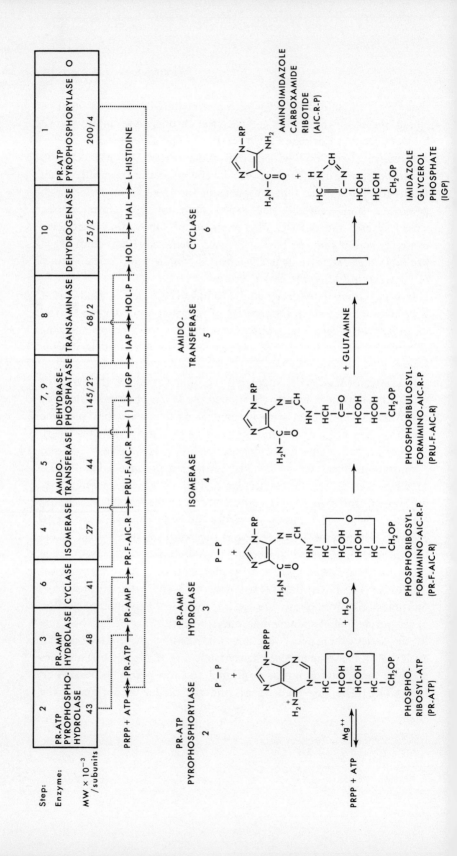

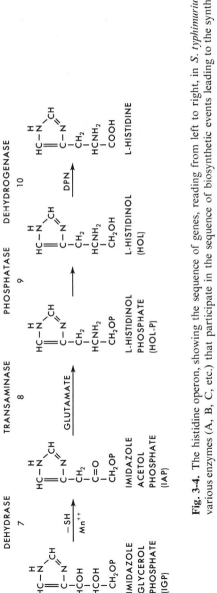

Fig. 3-4. The histidine operon, showing the sequence of genes, reading from left to right, in *S. typhimurium* coding for the various enzymes (A, B, C, etc.) that participate in the sequence of biosynthetic events leading to the synthesis of L-histidine (from Loper et al., 1964).

purine ring becomes detached as aminoimidazole carboxamide ribotide and leaves the histidine pathway. This compound is the precursor of the purine ring in purine biosynthesis. The remainder of the molecule of PR-ATP condenses to imidazole glycerol phosphate (step 5, Fig. 3-4). The side chain is oxidized, transaminated, dephosphorylated, and oxidized to histidine, which thus is formed as a degradative product of adenine. Surprisingly, the imidazole group of adenine is discarded in the process even though histidine is itself an imidazole. The feat of mapping the histidine gene cluster was carried out by Hartman, Ames, and their collaborators (Loper et al., 1964; Benzinger and Hartman, 1962; Ames and Garry, 1959), who isolated and examined about 900 histidine-requiring mutants in the process and found that each mutation was located within a short segment, estimated at 13,000 base pairs, of the bacterial chromosome. About 540 of the mutations were mapped in the sequence, which contains nine genes, each concerned with the synthesis of an enzyme. One of these enzymes, dehydrasephosphatase, carries out two enzymatic steps, and another, histidine dehydrogenase, performs a double oxidation reaction of histidinol to histidine through the aldehyde. The synthesis of the long messenger RNA molecule which codes for all nine enzymes is subject to *coordinate repression*, a term meaning that the production of the entire RNA molecule may be turned on and off by an operator gene, which is placed immediately to the right of the sequence and is shown as "O" in Fig. 3-4.

The amount of each enzyme synthesized varies over a wide range and is subject to control of the production of the long "polycistronic" messenger by the operator gene. The ratio of activity of one enzyme to another remains constant, so that the rate of production of each intermediate substance in the series of reactions stays in balance, and excessive amounts of any one of them do not accumulate except in mutants in which the balance is upset by blocking one of the enzymatic steps (Loper et al., 1964).

A theory of "modulating" triplets marking the junction between each successive pair of enzymes was advanced by Ames and Hartman (1963). This process depends upon a chain-releasing triplet occurring at the end of each segment of messenger RNA which codes for a single enzyme and the existence of a modulating triplet or triplets at the beginning of the next segment. The effect of modulation would be to slow down the rate of translation of each successive gene, so that less and less of each enzyme would be made as the message travels over the ribosomes. Enzymes coded by the distal end of the operon would therefore be made in smaller quantities than those at the right (operator) end.

The fact that the genes in the histidine operon are in close juxtaposition is of much evolutionary interest in the light of theories formulated by Horowitz and Lewis. The proposal by Horowitz (1945) relative to the evolution of biosynthetic pathways controlled by a group of functionally related genes has been widely quoted. The theory of the origin of life formulated by Oparin states that the atmosphere of the primeval earth was in a reduced state and contained large quantities of hydrocarbons and ammonia (Chapter 2). These raw materials gave rise to various organic compounds that reached high concentrations in the early oceans, thus supplying the nutritional requirements for the completely heterotrophic organisms that were the only form of life (Haldane, 1932). Horowitz' viewpoint was that growth of these organisms would use up the supply of essential food substances, thus creating what were presumably the first nutritional deficiencies in prehistory. The culture solution still, however, contained substances that were closely related to these essential nutrients. The presence of such substances led to the survival of organisms that were able to convert them to necessary metabolites; for example, when the supply of histidine became depleted, one would postulate that it was necessary to use histidinol (Fig. 3-4) as a precursor for histidine, and when the lysine supply became exhausted, organisms survived that were able to obtain their supply of lysine by the decarboxylation of diaminopimelic acid, a process that still exists in green plants (Chapter 8).

The depletion of histidinol and diaminopimelic acid would in turn lead to the appearance of the next backward step in the respective biosynthetic pathways. These successive steps were made possible by gene duplication followed by functional differentiation as suggested by Lewis (1951). A circumstance that supports these proposals is found in the fact that the genes responsible for certain metabolic pathways exist in clusters in the chromosomes of *E. coli* and *Salmonella*. Further support would be found if the series of enzymes concerned in the synthesis of metabolites such as histidine could be shown to be related in terms of their amino acid sequences in a manner similar to the hemoglobins. Horowitz (1965) has pointed out that *feedback inhibition*, in which the production of the last enzyme in the metabolic pathway inhibits the activity of the first enzyme, may indicate that the first enzyme carries a "memory" of its origin. Perhaps such a "memory" would be indicated by similar or analogous amino acid sequences occurring in the first and last enzymes of the series coded by the operon.

The genes that govern the synthesis of a sequence of enzymes in a metabolic pathway are clustered in the his operon in *Salmonella*, as noted. The genes controlling such sequences are, however, more usually scattered in other examples, and this may be due to changes such as recombination and crossing over, leading to relocation in different parts of the chromosome. An actual instance of such reshuffling was described (Pittard, Loutit, and Adelberg, 1963; Pittard and Ramakrishnan, 1964) in the case of a strain of *E. coli* (AB 1206) which had mutated from the wild type to produce a rearrangement in the arginine gene cluster. The synthetic pathway for arginine in *E. coli*, strain W, contains eight enzymatic steps (Vogel, Bacon, and Baich, 1963):

glutamic acid $\xrightarrow{1}$ N-acetyl glutamic acid $\xrightarrow{2}$ N-acetylglutamyl phosphate $\xrightarrow{3}$ N-acetyl glutamic acid γ semialdehyde $\xrightarrow{4}$ N-acetylornithine $\xrightarrow{5}$ ornithine $\xrightarrow{6}$ citrulline $\xrightarrow{7}$ argininosuccinic acid $\xrightarrow{8}$ arginine

The genes for (2) N-acetyl γ glutaminokinase: (3) N-acetyl glutamic semialdehyde dehydrogenase: (5) acetylornithase; and (8) argininosuccinase are clustered, but the other genes and the regulatory, or "operator," gene are scattered through the chromosome. The regulatory gene is controlled by arginine and under its influence *represses* the production of the enzymes no matter whether they are governed by scattered or clustered genes (Vogel et al., 1963). This may indicate that each gene has a repressor recognition site and that the repressor substance is produced by the regulatory gene, following which it is released and travels to the several repressor recognition sites. It was further noted by Baumberg, Bacon, and Vogel (1965) that the responses of "normal" *E. coli* and AB 1206 to repression by arginine were identical as measured by the rates of formation of acetylornithase and argininosuccinase. Therefore, the rearrangement of the genes for steps 5 and 8 did not affect the manner of reading the messenger in a way that would have been predicted by the modulator triplet proposal of Ames and Hartman (1963).

[4]

Mutations

Nothing of him that doth fade
But doth suffer a sea-change
Into something rich and strange.
Shakespeare, The Tempest I, ii. 394

Many scientific investigations and the conclusions arising from them have served during the past hundred years to enlarge and substantiate the concept of evolution. Three essential components, all of which are currently being explored in terms of molecules and chemical reactions, are in the concept. The first of these is the transmission and expression of genetic information; this takes place by the replication of DNA, by its transcription into messenger, transfer, and ribosomal RNA, and by the synthesis of proteins on the ribosomes. The second component is an incessant procession of changes in the individual bases in the long molecules of DNA, together with other alterations in these molecules, including crossing over, lengthening, deletion, and end-to-end duplication. Any of these changes can produce hereditary alterations. The third is the principle of the survival of the fittest, the process by which advantageous characteristics—the synthesis of a respiratory pigment, the fixation of nitrogen, the conversion of sunlight into chemical energy, a means of locomotion—enable their owners to thrive and reproduce. Such characteristics result from a complex web of biochemical reactions which emanate from proteins that are coded by the genetic message and that are subject to changes caused by mutations and other rearrangements of the DNA.

Within this framework of chemical evolution we perceive ourselves, the human species, driven onward by the same biochemical forces that induce other successful organisms to devour others or thrust them aside in the struggle for free energy and impelled by these same forces to examine and understand our environment so that we may alter it for our own purposes.

The term *mutation* implies a change. Its use is mentioned by Simpson (1953, p. 81) as having been first applied by Waagen in 1868 to "a recognizable stage in a continuously evolving lineage." This application differs from its present meaning, which relates to the sudden appearance of individuals with characters differing from those of their ancestry, these characters being transmissible to succeeding generations by normal hereditary processes.

At first sight this definition appears to be at odds with the principle which states that acquired characters are not hereditary, but this anomaly is nicely circumvented by attributing mutations to a change in the chromosomal material that is responsible for hereditary characteristics. This concept is substantiated by an enormous number of experimental observations.

It was found by Muller (1927) that X-rays would produce mutations in *Drosophila melanogaster*. Many of the mutations were similar to those which had been previously obtained spontaneously at much lower rate; others were new either with respect to morphology or to chromosomal location. An effect of X-rays in changing the chemical composition of the genetic material was thus demonstrated. It is worth noting that Muller's article (1927) warned against the indiscriminate use of X-rays in medical practice.

Subsequent investigations showed that certain chemical substances would produce mutations in *Drosophila*, including the alkylating agents nitrogen and sulfur "mustards" (Auerbach and Robson, 1946), formaldehyde (Rapoport, 1946), ethyl ethane sulfonate (Rapoport, 1947), and many others. These findings prepared the ground for later studies on the chemical basis of mutagenesis. The studies were greatly accelerated by the use of microbial systems, especially those involving bacteriophages (Benzer and Freese, 1958; Freese, 1959*a*), which can be studied in such enormous numbers in laboratory experiments. Indeed, Muller

102 Mutations

(1922) pointed out that

if these d'Herelle bodies were really genes, fundamentally like our chromosome genes, they would give us an utterly new angle from which to attack the gene problem. They are filterable, to some extent isolable, can be handled in test tubes, and their properties, as shown by their effects on the bacteria, can then be studied after treatment.

This approach, the use of d'Herelle bodies (bacteriophages), is now basic to the study of chemistry of genetics.

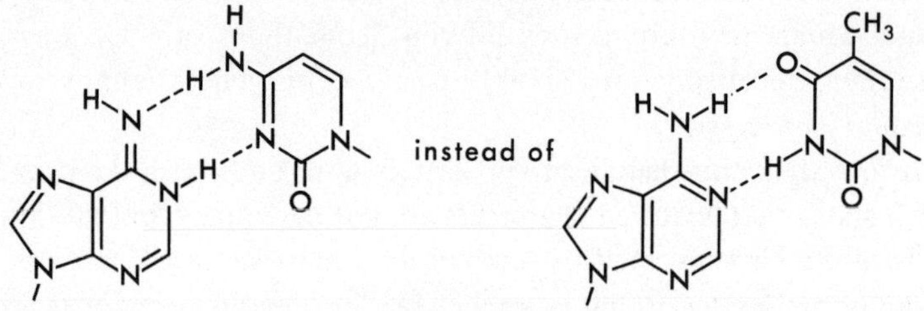

Fig. 4-1. "Mispairing" by adenine.

A proposal describing a molecular mechanism for mutations was made by Watson and Crick (1953c). Their explanation relates to *point mutations*—changes that occur at a single genetic locus, which is now known to be a pair of bases in the twin DNA strand. They pointed out that each of the four bases in DNA could shift rarely and occasionally to tautomeric modifications that were different from the usual form. The tautomer would pair with the "wrong" complementary base during replication of the DNA molecule. For example, the tautomeric form of adenine would pair with the "regular" form of cytosine instead of with thymine, as shown in Fig. 4-1.

Subsequent replications taking place in the normal manner without further similar errors would result in the replacement of an AT pair in the double-stranded DNA molecule by GC. This could lead to a change of a single base in a coding triplet and, as a result, to the substitution of one amino acid by another in a polypeptide chain. In turn, the substitution would under certain circumstances produce a biologically important change in the properties of the protein, leading to a phenotypic change. If this series of events took place in a single-celled organism, or in the ovum or spermatozoon of a multicellular organism, the offspring could be a discernible mutant.

When reinterpreted in terms of chemical mutagens, this disarmingly simple explanation provides a satisfactory model for many chemically induced and some spontaneous point-mutational phenomena. Examples will be given in detail later in this chapter.

The tautomerically induced mispairing of bases is one of many possibilities for producing changes in the sequence of base pairs in DNA. Most of the investigations in this field have been carried out with bacteria and bacteriophages, especially bacteriophage T4. In these studies, many different physical and chemical stimuli were found to be mutagenic owing to their effect on the base-pairing mechanism, including X-rays, ultraviolet light, hydrogen ions, nitrous acid, 5-bromouracil, certain alkylating agents, acridines, hydroxylamine, and heat. The effect of nitrous acid can readily be visualized: it deaminates cytosine to uracil, which pairs with adenine instead of with guanine, adenine to hypoxanthine, which pairs with guanine instead of with thymine, and guanine to xanthine (Vielmetter and Schuster, 1960). The action of nitrous acid should therefore be similar to that of reagents such as hydrogen ions, which could produce mispairing because of tautomeric shifts. In both cases, the switch should be the substitution in DNA of one purine by another (pu/pu) or one pyrimidine by another (py/py). Freese (1959*a*) gave the name *transitions* to these interchanges. A transition produced by nitrous acid is shown in Fig. 4-2.

The other type of single-base changes possible in DNA was named *transversions* (Freese, 1963); these are pu/py or py/pu interchanges. They are not produced by mispairing or deamination in the manner described above but would instead appear to be the result of a more drastic change in a single DNA strand so that perhaps a purine is inserted in place of a pyrimidine (or vice versa) owing to a local breakdown of complementary pairing. In subsequent replications a normal pairing would take place so that the following possibilities would exist

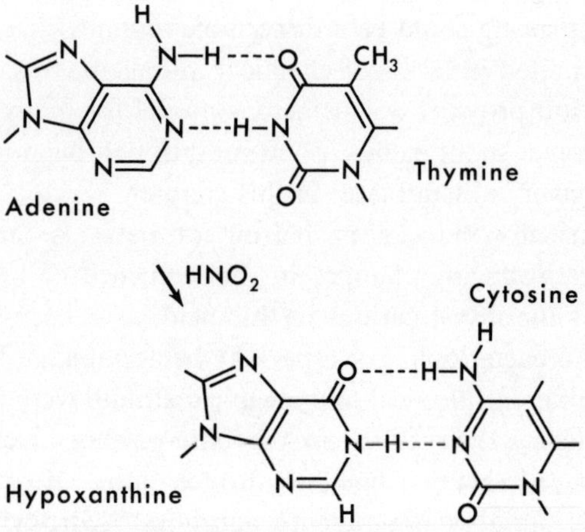

Fig. 4-2. Production of a transition by HNO_2.

for the replacements:

$$\frac{A}{T} \to \frac{C}{G} \text{ or } \frac{T}{A} \qquad \frac{T}{A} \to \frac{G}{C} \text{ or } \frac{A}{T} \qquad \frac{C}{G} \to \frac{A}{T} \text{ or } \frac{G}{C} \qquad \frac{G}{C} \to \frac{T}{A} \text{ or } \frac{C}{G}$$

It should be noted with respect to transitions that $A \to G$ and $G \to A$ in messenger RNA can *both* result from deaminative changes in DNA as follows:

$$\begin{array}{lll}
\text{DNA strand 1} & \ldots A \ldots & \xrightarrow{HNO_2} \ldots H \ldots \\
\text{DNA strand 2} & \ldots T \ldots & \phantom{\xrightarrow{HNO_2}} \ldots C \ldots \\
\text{Messenger RNA} & \ldots \overline{A} \ldots & \phantom{\xrightarrow{HNO_2}} \ldots \overline{G} \ldots
\end{array}$$

or

$$\begin{array}{lll}
\text{DNA strand 1} & \ldots G \ldots & \xrightarrow{HNO_2} \ldots A \ldots \\
\text{DNA strand 2} & \ldots C \ldots & \phantom{\xrightarrow{HNO_2}} \ldots U \ldots \\
\text{Messenger RNA} & \ldots \overline{G} \ldots & \phantom{\xrightarrow{HNO_2}} \ldots \overline{A} \ldots
\end{array}$$

Similarly, $C \to U$ and $U \to C$ in messenger RNA can result from

$$\frac{\begin{array}{c}C\\G\end{array}}{C} \xrightarrow{HNO_2} \frac{\begin{array}{c}U\\A\end{array}}{U} \quad \text{and} \quad \frac{\begin{array}{c}T\\A\end{array}}{U} \xrightarrow{HNO_2} \frac{\begin{array}{c}C\\H\end{array}}{C}.$$

Changes in either strand of the DNA can be mutationally significant in relation to the genetic code, since a single-base change in one strand will produce a complementary change in the other strand at replication.

Mutations may also be considered in terms of amino acid changes and coding triplet relationships. Let us consider the triplet GCU as a code for alanine. Single-base transitional changes can produce only three different triplets, ACU, GCC, and GUU. If, however, all three of the bases in the triplet undergo transitions, the total number of possible different triplets formed from GCU will be seven. Alanine is considered to be coded by GCA, GCG, and GCC as well as by GCU. The total number of different triplets that can possibly be formed by transitions from these three triplets is 14, as follows: ACA, GUA, GCG, AUA, ACG, GUG, AUG, ACC, GUC, GCU, AUC, ACU, GUU, and AUU. Some of them duplicate each other as different codes for the same amino acid.

Alanine is a common amino acid and occurs at many sites in hemoglobin and cytochrome c. It is often replaced by other amino acids at homologous sites in the various hemoglobins and cytochromes c. No less than 18 different amino acids, all the amino acids except tryptophan, replace alanine at such sites.

In terms of the triplets in Table 2-12, the amino acids that are related to the ala codes GCU, GCC, GCA, and GCG by transitional mutations are ilu, met, thr, and val. In terms of both transitional and transversional changes, the relationships and the incidence of these changes in the hemoglobin and cytochrome sites are shown in Table 4-1. Single-base changes outnumber two- and three-base changes by 2 to 1. Evidently some of the interchanges involving alanine in the hemoglobins and cytochromes c are due to transversions.

In a living organism, the interchange of amino acids in a protein is dependent upon the acceptability of the interchanges in terms of their effect on the structure and the properties of the protein. Exchanges at some loci may take place freely; for example, in the various cytochromes c, seven amino acids are found at residue 89 (Chapter 6). In contrast, residue 91 is occupied solely by arginine. Asparagine interchanges with four other amino acids at sites 92 and 103, but is unchanged in the cytochromes *c* of 14 different species at sites 31, 53, and 70. It appears that various sites in the same protein may be either highly receptive, moderately receptive, or unreceptive to mutational changes. In some cases, it has been found that the mutational substitution of one amino acid by another may permit the protein to function partially,

Table 4-1. Occurrences of Single-Base and Two-Base Changes from Alanine in Hemoglobins and Cytochrome *c*

Single-Base Changes		Two-Base Changes	
Amino Acid	*No. of Occurrences*	*Amino Acid*	*No. of Occurrences*
asp	11	arg	2
glu	15	asN	5
gly	18	cys	2
pro	9	glN	3
ser	21	his	3
thr	13	ilu	7
val	13	leu	14
		lys	11
		met	2
		phe	2
		tyr	1
		try	0
Total	100		52

although not with full efficiency. An example of this is a partial revertant of a mutation in tryptophan synthetase, a protein described by Yanofsky (1964). Locus 8 in peptide CP2 contains glycine in the wild-type protein. A mutant (A46) in which this glycine in the A protein is replaced by glutamic acid is enzymatically inactive. A46 gives rise to a second mutant, containing valine instead of glutamic acid at the above locus, and this protein is partially active.

Yanofsky also reported that partial activity could be produced in A46 by a mutation producing an amino acid substitution at a *different* site. This mutant (A46-PR8) retained glutamic acid in peptide CP2, but had cysteine replacing a tyrosine residue in another peptide. A similar effect was described earlier for the alkaline phosphatase of

E. coli by Garen, Levinthal, and Rothman (1961*a*). This phenomenon appears to be due to the effect of different segments of a protein chain upon each other. It has also been suggested that suppressor mutations may also play a part in such phenomena by changing the translation of coding triplets.

Chemical Mutagenesis

The experimental production of mutations with specific agents is detected by the appearance of changes in the progeny or by lethal effects. One of the earliest findings in this field, as noted above, was Muller's observation that X-rays greatly increased the frequency of lethal or other mutations in *Drosophila melanogaster*. The relation between X-ray dosage and percentage of mutations in the X chromosome of *Drosophila* is approximately linear. The findings with X-rays were followed in subsequent experiments that demonstrated the effect of mutagenic chemicals on bacteria and bacteriophages. In such experiments, the primary action of the mutagen is on the DNA. This action is reflected in altered properties of the organism but is not measurable with available procedures in terms of localization of and description of the chemical changes in DNA. As will be discussed below, inferences as to the base changes in DNA caused by mutagens may be drawn by examination of reversions of the mutations that are in many cases produced by mutagens whose specific chemical effect on the different bases has been determined by separate studies.

The use of chemical mutagens in the study of mutations has found its greatest scope in experiments with bacterial viruses, which can be examined rapidly in large numbers. They reveal phenotypic changes, such as *rapid lysis* (r^+), by modifications of the appearance of the plaques found on bacterial culture plates. The mutagens that were first studied produced no effects on extracellular bacteriophages, but mutations were produced in virus-infected bacteria. Bromouracil (BU) was discovered by Litman and Pardee (1956) to be very active in this respect, and 2-aminopurine was later shown to be similarly effective (Freese, 1959*a*). These mutagens can be inserted in the developing DNA chain by enzymatic reactions taking place in the bacterial cell.

It was subsequently found that certain other mutagens were active on extracellular bacteriophages, including ethyl methane sulfonate, hydroxylamine, and nitrous acid, which may be visualized as reacting with bases that were previously incorporated in the DNA chain, so that pairing errors are subsequently committed by the altered base.

The incorporation of BU in place of T into DNA of bacteria and bacteriophage was demonstrated by Dunn and Smith (1954). BU differs from T in that the electronegativity of Br is higher than that of the methyl group in T. As a consequence, the tautomeric modification of the pyrimidine ring occurs more frequently in BU than in T and mispairing of BU with G is encouraged during the incorporation of BU, and the mutation rate will be increased by growing the organism in the presence of BU owing to *incorporation errors*. Once incorporated, either in a mutant or in a nonmutant, BU will usually pair with A. The net result is a transition from GC to AT.

The incorporated BU, because of suppression of ionization by polymerization, has less tendency than free BU to make pairing errors. Replication errors will occasionally occur, however, resulting in BU pairing with G, thus restoring the GC pair at the original locus. This explanation could provide for forward and reverse mutations.

Further refinements in the study of transitions in chemical mutagenesis are provided by using different mutagens to produce reversions. Champe and Benzer (1962a) studied mutants of the r-II region in bacteriophage T4 that were inducible to reversion by the base analog 2-aminopurine, which produces transitions (Freese, 1963a). Of 69 such mutants treated with BU deoxyriboside and hydroxylamine, 16 were found to be readily reversible by these two mutagens, leading to the conclusion that the reversions were due to transitions of GC to AT.

As an extension of these observations, Champe and Benzer infected *E. coli*, strain K, with bacteriophage mutants and grew the infected cells in media with a base analog mutagen 5-fluorouracil (FU), which is incorporated into messenger RNA in place of uracil (U). It was found that FU partially reversed the defective phenotypes of certain of the mutants, leading to the conclusion that FU, after incubation so that it was introduced at the U sites in the messenger RNA, occasionally behaved like cytosine and corrected the coding error (C → U) at the site of protein synthesis.

Benzer's work shows the length to which mutations may be carried as a means of investigating the base composition of genetic DNA. The studies were carried out without knowledge of the function or amino acid sequence of the protein coded by the region under investigation. Benzer's results showed that some sites in the A and B genes, or cistrons, of the r-II region in T4 bacteriophage are far more susceptible than others to spontaneous or chemically induced mutations. A complex pattern was found in the results; the total number of mutations at each site was not the same for each mutagen, and one site in particular (site 117 in segment B4b1) mutated far more often spontaneously than in response to mutagens. Some sites, including site 117, often mutated spontaneously, others a few times, and others only once, and it was concluded from genetic analysis that there were yet other sites that did not mutate.

On a more restricted scale, an analogous situation exists in the extent to which different sites vary in the hemoglobins and cytochrome *c*. Here the explanation has been based on the relation of amino acid changes to the function of the proteins. Certain loci, such as 89 in cytochrome *c*, exhibit many different amino acid changes; presumably such sites can undergo a variety of changes without disturbing the secondary and tertiary structure of the protein to an extent that interferes with its biological activity. Other sites are relatively constant or are not known to change at all. The explanation in the case of bacteriophage is different, because mutants of T4 phage with a nonfunctional r-II region can be grown on *E. coli* strain B, and the damage is detected only by the fact that the mutants will not grow on strain K. The "hot spots" in this case are probably due to the type and arrangement of neighboring base pairs in the DNA.

In studies of bacteriophage T4, Freese (1959*b*) distinguished between two types of mutations as judged phenotypically by plaque types and growth characteristics in the *E. coli* strains B and K. The first type included those induced by the base analog mutagens 2-aminopurine

(Ap) and 5-bromodeoxyuridine (BUdR). Mutants of phage T4 induced by these base analogs in the forward direction could be induced to revert. The sequence of events in the series wild type → mutant → revertant can be *two successive transitions* A → G → A or T → hydroxymethylcytosine (HC) → T, at the same locus. This mutant → revertant effect was observed, and was particularly striking when bromouracil mutants were treated with AP, the rate of reversion being increased as much as 10^5 times.

The second type of mutation, found in spontaneous mutants, was, with one exception, not inducible to reversion by treatment with base analogs. Freese concluded that these mutants represented *transversions*, since the substitution of one purine by another purine or one pyrimidine by another pyrimidine caused by the base analog would not restore the original base to the changed locus. For example, a change of T to A would not be reverted by a subsequent change of A to G. Five mutations of this type were noted in the series of experiments described by Freese; a sixth was induced to revert by AP and BUdR and was hence presumably due to a transition.

Freese also examined 40 mutants induced by proflavine (3,6-diaminoacridine) mutants. Only one, a single occurrence, was inducible to revert by base analogs, although the mutants reverted spontaneously. He suggested that proflavine might induce transversions, in view of the nonrevertibility of proflavine mutants by base analogs.

An alternate explanation for the effect of proflavine and its congeners was offered by Brenner et al. (1961), who took issue with Freese's proposal that proflavine produced transversions and suggested that it actually induced deletions and additions of base pairs. Brenner and his coworkers supported the argument by proposing that the proflavine compounds became intercalated in the DNA strand, thus allowing the interposition or deletion of a base pair and changing the reading frame of the sequence of triplets in the code. They also suggested that revertants of proflavine-type mutants may not actually be due to restoration of the original base pair at the mutated site but instead to a second mutation, in which the second change takes place at a different location from the first but nevertheless restores the function of the organism (the phage). In terms of the base sequence in DNA, the forward mutation is represented as the deletion or addition of a single base. This displaces the orderly sequence of coding triplets that lie on the right of (distal to) the altered site, continuing to the end of the structural gene, so that a partially completed or inactive protein molecule is produced. The backward mutation is visualized as the addition or deletion of another single base at a nearby site, thus correcting the displacement except for a short region between the two changed sites. The change in this region is evidently not critical; indeed, Crick et al. (1961) have concluded that the part of the protein coded by the B1 segment of the B cistron of the r-II region of bacteriophage T4 is probably not essential for its function.

The single-strand DNA bacteriophages S13 and ϕX 174 were used by Tessman, Poddar, and Kumar (1964) and by Howard and Tessman (1964a, b) in studies of chemical mutagenesis leading towards the identification of bases at the mutated sites. The use of single-stranded rather than double-stranded bacteriophages simplified the problem of identifying a mutagenically changed base at a given site. Forward and reverse mutations were examined.

Let us suppose that mutagen I changes A to G, mutagen II changes C to T, mutagen III changes G to A, and mutagen IV changes T to C. Mutants produced by I will be reverted by III but not by I, II, or IV, and so on. Then if one of the changes produced by I and III is known and one of the changes produced by II and IV is known, a site giving rise to a point mutation can be identified.

Tessman et al. (1964) used hydroxylamine (HA), ethyl methane sulfonate (EMS), and nitrous acid (NA) as "external" mutagens. They treated the bacteriophages in vitro in lysates and plated the resultant products on agar containing appropriate hosts.

In a series of experiments with S13, 114 HA-induced mutants were isolated, and it was possible to identify 112 distinct mutant types from the appearance of the plaques when the phages grew on a variety of sensitive hosts. It was shown by genetic analysis that the mutants occupied 112 different genetic sites, two of which were each occupied by two mutants, which in each case appeared to be identical. These experiments showed that different phenotypes were produced consistently by different genotypes.

The experiments with the three mutagens resulted in specific responses, particularly when both forward and reverse mutations were studied, indicating that only one species of base was changed for each mutation.

A problem that arises in studying forward and reverse mutations is that the reverse mutation may be a change at a different site, which

Table 4-2. Induction of S13 Mutations by Hydroxylamine (HA), Ethane Methane Sulfonate (EMS), and Nitrous Acid (NA)

Obser-		Induction Frequency[a]						Inferred Base Change in Forward Mutation
vation	Mutation	HA		EMS		NA		
		F	R	F	R	F	R	
1	$h^+ \to h_i 1$	0	++	+	++	+	++	T → C
2	$h^+ \to h_i 2$	0	0	0	++	++	++	A → G
3	$h^+ \to h_i 1$	0	0	++	0	++	++	G → A
4	$h^+ \to h_i 2$	0	0	++	+	++	++	G → A
5	$h^+ \to h_i 65$	0	0	++	0	++	++	G → A
6	$h_i UR48 \to h_i UR48S$	++	...[b]	++	...[b]	++	...[b]	C → T

F = forward mutation; R = reverse mutation; ++ = high frequency (>10⁻⁶); + = low frequency (<10⁶); 0 = no detectable induction.
[a] Number of mutants per survivor per phage lethal hit.
[b] No results available.
Source: Tessman et al. (1964), which contains other details.

simulates a reversal of the original change, as discussed above. The chances of such an occurrence accounting for the results were greatly reduced by carrying out the forward-reversal procedure serially in certain cases for as many as 12 steps.

The pattern of mutagenesis for S13 mutations is shown in Table 4-2. It was concluded from the results of other investigators that HA produces the reaction C → T, and it follows that observations 2 to 5 in Table 4-2, in which HA had no effect, involved interchanges of A and G and that observation 1 represents T → C in the forward mutation. The second conclusion, made tentatively, was that EMS induced G → A strongly and A → G only slightly. Surprisingly, NA induced all four possible transitional changes. It had previously been concluded that the deamination of G to produce xanthine (X) would be lethal because xanthine deoxyriboside triphosphate does not participate in the enzymatic synthesis of DNA (Bessman et al., 1958). If, however, the deamination of guanine to xanthine takes place in the DNA chain, this condition does not arise, and the xanthine might pair with T (Tessman et al., 1964; Lawley and Brooks, 1962; Cavalieri et al., 1954).

Howard and Tessman (1964a) treated S13 with 5-bromodeoxy uridine (BUdR) and 2-aminopurine in vivo, i.e., in bacterial cells that had been exposed to the mutagens. This procedure enables BU to be substituted for T in the DNA chain during formation of the double-stranded replicative form of the bacteriophage. As noted previously, the inclusion of BU in the chain, replacing T, induces mutations because mispairing with G rather than A occurs more frequently with BU than with T. The occurrence of such mispairing should be reduced by adding thymine (or thymidine) to the growth medium to compete with BU for incorporation into T sites; and this result was found, indicating that replication of S13 is accompanied by pairing in vivo with a complementary strand. Furthermore, the mutations that were inferred by Tessman et al. (1964) to have been T $\rightarrow$ C or A $\rightarrow$ G changes (see Table 4-2) were induced by BUdR and repressed by thymidine. Those that had been inferred to be C $\rightarrow$ T or A $\rightarrow$ G were not induced by BUdR, with the exception of one that was induced slightly and was repressed by cytosine deoxyriboside rather than by thymidine. These results substantiated the conclusions of Tessman et al. (1964). Howard and Tessman (1964b) found that S13 mutants induced by ultraviolet light showed a reversion pattern with chemical mutagens, indicating that 11 of 16 mutants were caused by C $\rightarrow$ T changes. In the remaining 5, transitional changes other than C $\rightarrow$ T were excluded. In an earlier publication, Drake (1963) studied the properties of UV-induced mutants of the r-II region of the bacteriophage T4 and examined their reversibility by UV, 2-aminopurine, 5-bromouracil, hydroxylamine, and proflavin. About half of the mutants were induced to revert by the base analogs in a manner indicating that an A-T base pair at the mutant site was being reconverted to G-HMC. Most of the others were reversible by proflavine.

These results are reviewed in some detail because they show the extent to which the base composition may be explored by specialized techniques applied to a "living" DNA strand. The degree of analytical precision achieved by Tessman represents a high refinement of experimental genetic techniques. It is due to (1) the development of the genetics of bacteriophage; (2) the methods developed by Benzer, who refined genetic analysis to the recognition of a single base pair; (3) the use of single-stranded bacterial viruses, which are the smallest and simplest known living forms and which have the lowest DNA content; and (4) inferences drawn from experiments on chemical effects of specific mutagens on purine and pyrimidine bases. The results indicate the scattering of mutational changes along the DNA molecule.

Tessman's findings, extending beyond those of Champe and Benzer (1962a), indicate that transitional base changes may be identified at point mutations. The current views concerning gene-protein relationships are supported by such results; chemically induced mutations which produce amino acid changes in tobacco mosaic virus (TMV) coat protein (Tsugita and Fraenkel-Conrat, 1960; Funatsu and Fraenkel-Conrat, 1964; Wittmann and Wittmann-Liebold, 1963) are attributed to base changes in TMV RNA, although the actual base changes in the RNA have not been demonstrated. The studies with bacteriophages show that such base changes do indeed take place in a manner which is consistent with single-amino-acid changes in proteins coded by a strand of nucleic acid. The studies also indicate that all four possible types of transitions may be caused by nitrous acid. This finding agrees with the coding relationships seen in the amino acid substitutions induced by nitrous acid in TMV. The results with TMV further resemble the findings with bacteriophage in showing that certain loci are more prone than others to mutations (Wittmann and Wittmann-Liebold, 1963). In the case of T4 bacteriophage, certain sites are highly susceptible to mutation with bromouracil, and still other sites show high susceptibility to mutation with 2-aminopurine. Furthermore, the in vitro mutagens, ethyl methane sulfonate and nitrous acid, produce a topographic map of mutations that differs again from the spontaneous, BU, and AP maps. One must conclude that different mutagens can behave differently towards various configurations that occur at specific regions in the DNA chain, but not enough results with TMV are available to examine the possibility of a similar conclusion. There is some evidence that mutations do not appear in certain regions of the amino acid sequence of TMV coat protein and hence are not tolerated (*ibid.*).

The use of bacteriophage T4 in studies of mutations was now to move on to its greatest deductive triumph in the hands of Brenner and his associates, who brought chemical mutagenesis and amino acid sequence studies to bear on the problem of the "nonsense" triplets.

The term *nonsense triplet* is the product of a neology which at times has an obfuscatory effect on the uninitiated. We shall endeavor to clear the air by anticipating our story; the term is a misnomer for a *chain-terminating triplet*. The final amino acid in a polypeptide chain must be released from ester linkage with sRNA by hydrolysis to liberate the α-carboxyl group. The mechanism of the hydrolytic process is unknown, but it is preceded by a step in which the addition of amino acids to the growing polypeptide is halted. The coding of this step, which is anything but nonsensical, is the function of a chain-terminating triplet in messenger RNA. In Table 2-12 these triplets were stated to code for "gaps." A mutation which produces such a gap at the wrong place in a protein is termed a *nonsense mutation* because it prematurely ends the polypeptide chain (Garen and Siddiqi, 1962). There is no evidence that polypeptide synthesis is reinitiated on the distal side of the chain-terminating triplet to complete the balance of the polypeptide in such mutants.

Here another term must be introduced, *suppression*. The term is used to refer to a mutation at a genetic site which "suppresses" the effect of a previous mutation at a different site. The result is that the mutant phenotype caused by the first mutation is restored to its original wild type. One change which would have the effect of suppression consists of the addition of a base pair to a gene at a short distance distal to a locus from which a base pair has been deleted, or, conversely, the removal of a base pair a few residues distal to an addition, thus restoring the "reading frame" of the code.

A second mechanism of suppression, demonstrated by Brenner et al. (1965) and by Weigert and Garen (1965a), is exemplified by the insertion of serine at a UAG messenger triplet locus, thus repairing an interrupted protein chain. A better word than "suppression" in this example might be "repair," since the effect of a chain-terminating mutation on a protein is restored or repaired by a second mutation elsewhere in the genome that enables the chain-terminating triplet to incorporate an amino acid instead of breaking the chain. In such a case it would appear that the chain-terminating triplet was misread in terms of the amino acid code owing to the production of an unusual sRNA by the suppressed mutant.

The r-II gene of bacteriophage T4 actually consists of two genes, the A and B cistrons in Benzer's terminology, with respective sequences of about 1,700 and 1,100 nucleotide pairs. The two proteins coded by the A and B cistrons are needed for growth in *E. coli* strain K, which is a lysogenic strain carrying the prophage of λ bacteriophage. However, bacteriophage T4 can grow on *E. coli* strain B, the wild type, even if there are defective mutations in the r-II A and B cistrons. Perhaps the genome of *E. coli* strain B furnishes the enzymes which are also coded by normal A and B cistrons of T4.

A deletion mutation of r-II exists, termed r1589, which connects the A and B cistrons (Champe and Benzer, 1962b). Apparently the deletion includes the chain-terminating triplet that normally separates the proteins coded by A and B so that in r1589 a single "AB" protein is produced, shorter than normal A + B. This protein does not have the normal function of A but retains the function of B, as shown by the "cis-trans" test (Benzer and Champe, 1962). The insertion of additional mutations into the A cistron of r1589 was carried out. Some of these were found to be "nonsense" mutations, as shown by the fact that, although they occurred in A, they eliminated the function of B, and the mutants would hence not grow on *E. coli* strain KB. This would result from interruption of the translation of the genetic message in the A region of AB protein so that the B protein was not synthesized (Fig. 4-3). In addition, it was found that these mutants would grow on a suppressive host, *E. coli* KB3, which apparently supplied a modified sRNA or activating enzyme that brought an aminoacyl sRNA to the chain-terminating triplet.

At this point, a new term, *amber mutant*, must be introduced. It is defined (Epstein et al., 1963) as a mutant of bacteriophage T4D which forms plaques on *E. coli* CR 63 (a suppressor) but not on *E. coli* B. It was later shown that the "amber" mutation is to one of the chain-terminating triplets.

Sarabhai et al. (1964) studied amber mutants of the head protein gene, which is distinct and separate from the r-II gene, in T4. They

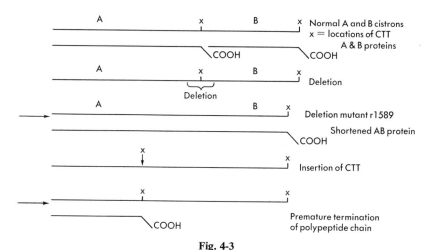

Fig. 4-3

were able to prepare sufficient quantities of head protein for fingerprinting its tryptic and chymotryptic peptides, so that mutations in bacteriophage were for the first time examined in terms of protein sequences. Ten different amber mutants were shown to interrupt the synthesis of head protein at 10 different and sequential points, and the collinearity of gene and protein was demonstrated by genetic mapping of the ten mutants.

Brenner et al. (1965) found that an amber mutant H36 of a head protein gene was caused by a mutation in the coding triplet for glutamine in a polypeptide of the head protein of the bacteriophage T4D. Evidently, the glutamine triplet had become changed into a chain-terminating triplet. Further studies showed that this triplet coded (or "miscoded") for serine when the mutant was grown on a suppressor strain of *E. coli* designated as su_1^+.

Attention was next given to amber mutants in another region of the T4 genome, the r-II gene in the mutant r1589 (Champe and Benzer, 1962*b*). In addition to the amber interruptions, it was found that a second type of mutation also could interrupt the synthesis of the polypeptide. This interruption was not restored on su_1^+ but was restored by other suppressor strains such as su_B^+. The second type of mutation which caused interruptions of the polypeptide chain was designated as an *ochre* triplet, to differentiate it from "amber," which is a translation of the name of Bernstein, a graduate student who collaborated in the discovery of mutants of this class at the California Institute of Technology (Edgar and Susman, 1962). The etymological significance of "ochre" is obscure (Brenner and Beckwith, 1965).

The r-II amber and ochre mutants were treated with chemical mutagens to produce reversions in which the reading of the polypeptide chain was restored. Reversions had been shown to be produced by 5-fluorouracil by Champe and Benzer (1962*a*), who concluded that this mutagen acted on DNA to produce a $U \rightarrow C$ effect in the messenger RNA. Hydroxylamine, which produces a $C \rightarrow U$ effect (Table 4-2) was ineffective. Their technique did not permit the bacteriophage to replicate, and so they detected changes only in the minus strand of DNA that participates in the RNA polymerase reaction and is complementary to messenger RNA. Brenner and coworkers used hydroxylamine and grew the treated bacteriophage on *E. coli* strain B, which permits replication of mutants in the absence of a functioning r-II cistron. The progeny were then plated on *E. coli* strain K to detect reversions. Under these conditions, hydroxylamine still did not produce reversions, thus showing that the corresponding messenger triplets either did not contain C or G or that a G or C had been converted into an A or T in a second chain-terminating triplet by the effect of the mutagen on a G:C or C:G base pair in the DNA. The mechanism of this will be shown later.

It was next found that ochre mutants could be converted into amber mutations by treating the mutants with 2-aminopurine or 5-bromouracil. This conversion was shown by a change in response to suppressors after treatment with these mutagens, both of which induce the transition $A:T \leftrightarrows G:C$. This mutation could not, however, be induced with hydroxylamine, which induces $G:C \to A:T$ but not $A:T \to G:C$. Therefore, the amber triplet contained G or C in place of A or U in the ochre triplet. Furthermore, amber and ochre mutants could both be induced from the wild-type triplet by hydroxylamine, showing that both amber and ochre had at least one A or T in common.

An experiment was then carried out to determine whether the amber triplet contained both U and A. Wild-type T4 r$^+$ bacteriophages were treated with hydroxylamine to induce mutations in the r-II gene. Hydroxylamine changes $G:C$ in the DNA to $G:U$, or $C:G$ to $U:G$. If the change is from $G:C$ to $G:U$ and the U is in the strand that codes the messenger, then the change would be immediately expressed without the DNA passing through a replication cycle. Such an immediate change would therefore indicate the presence of A in the amber triplet. If, in contrast, the change is from $C:G$ to $U:G$ and if the bacteriophage was allowed to pass through a replication cycle, the $U:G$ becomes $U:A$, and a delayed step would be produced in the messenger from C to U, thus indicating U in amber (Table 4-3). Set (2) mutants would be eliminated by growth on a nonsuppressive strain. They would appear only when the mutants were grown immediately on a suppressive strain. Set (1) mutants would survive on the nonsuppressive strain. They would therefore appear on a suppressive strain (su$^+$) after passage through a nonsuppressive (su$^-$) strain. It was found that both types of changes took place. The conclusion was that the amber triplet contained both A and U. Ochre triplets were also induced by the same procedure although not so strongly. This showed that the ochre triplet also contained an A and a U. The conclusion was that the amber triplet was either (UAG) or (UAC) and that the corresponding ochre triplet, induced by a deaminative transition, was either (UAA) or (UAU).

The sequence of the bases was not specified by these experiments but was deduced as follows: the amber triplet is connected by transitional mutations to the glutamine and ochre triplets. It was also found that amber mutants could be induced from a tryptophan locus by

Table 4-3. Deciphering the Amber and Ochre Triplets

(1)

DNA	C $\xrightarrow{NH_2OH}$ U		$\xrightarrow{replication}$	U	
	G	G		A	
mRNA	C wild	C		U	amber
		No change in mRNA		C/U change in mRNA $\therefore$ amber contains U	

(2)

DNA	G $\xrightarrow{NH_2OH}$ G		
	C	U	
mRNA	G wild	A	amber
		G/A change in mRNA $\therefore$ amber contains A	

Interpretations

(1)	CAG	$\rightarrow$	UAG	and	CAA	$\rightarrow$	UAA
	glN		amber		glN		ochre

(2)	UGG	$\rightarrow$	UAG	and	UGA	$\rightarrow$	UAA
	try		amber		try		ochre

Also	UAA $\xrightarrow{\text{2-amino purine or 5-bromouracil}}$ UAG,	but	UAA $\xrightarrow{NH_2OH}\!\!\!/\!/$ UAG
	ochre amber		ochre not to amber

Source: Brenner, Stretton, and Kaplan (1965).

2-aminopurine and 5-bromouracil. Since both tryptophan codes contain U and G (Table 2-11), this establishes the amber mutant as UAG, connected by transitional mutations with CAG, glN, UGG, try and UAA, ochre. These results agree with the findings by Weigert and Garen (Chapter 2).

Brenner et al. (1965) suggest a mechanism for polypeptide chain termination in which a "special" sRNA carries, in place of an amino acid, a chemical group that ruptures the ester bond between the α-carboxyl of the most recently added amino acid and the 3′-OH of ribose in the terminal adenosine of sRNA. This special sRNA, coded by UAA or UAG, is postulated to be the procedure for breaking the polypeptide chain at its correct termination point as specified by the genetic message. No "special" sRNA of this type has yet been discovered. A chain-terminating mechanism is, however, essential to life and is also of great evolutionary significance. An examination of Table 5-2 indicates that a mutation from GAe to UAe in the myoglobin gene may have been responsible for "lopping off" the six terminal acids in myoglobin when the primitive hemoglobin gene was formed.

It was concluded that UAA is more likely than UAG to be the "regular" chain-terminating triplet (Brenner et al., 1964). The reason for this was as follows: the amber mutants (UAG) are readily suppressed but the ochre mutants are suppressed only weakly. Strong suppression would tend to be lethal because it would join many proteins together by serine linkages. In contrast, weak suppression would result in most of the UAA triplets failing to incorporate amino acids, and this state of affairs would be necessary if the proteins of the organism are to be kept separate so that it can survive.

Mutations in Alkaline Phosphatase

This enzyme in *E. coli* has a molecular weight of 80,000 and is a dimer containing identical subunits and two active sites (Levinthal, Signer, and Fetherolf, 1962). Inactive forms of the enzyme can be produced by point mutations, presumably single-base changes, in *E. coli*. These mutants can revert to forms (pseudorevertants) which can produce altered but fully active enzyme proteins.

Certain phosphatase-negative mutants of *E. coli* have been studied that show an absence of any protein with the enzymatic or immunological properties of alkaline phosphatase. Fifteen such mutants, involving at least seven different sites in the gene, were described by Garen and Siddiqi (1962). These mutants were all suppressible by a gene that restored the enzyme to normal as measured by electrophoretic mobility, heat stability, and specific activity. The suppressed mutants produced from 3 to 100 percent of the normal amount of enzyme. It was stated that "nonsense" r-II mutants of bacteriophage T4 were restored to function by the same suppressor, following which Brenner et al. (1965) proposed that both types be termed amber mutants.

The nature of the phosphatase-negative mutant H12 was discovered by Weigert and Garen (1965a) by study of its reversion; they devised a remarkable technique. The mutant was seeded on an agar plate containing a medium responsive to phosphate-positive revertants. A drop of a mutagen (ethyl methane sulfonate or N-methyl-N'-nitro-N-nitrosoguanidine) was placed on the agar, and the plate was incubated for 3 days at 37 C. In other experiments, nitrous acid was used as a mutagen. The revertants were isolated, grown in pure culture, and used for the preparation of the enzyme, which was hydrolyzed enzymatically and fingerprinted. The revertants, some of which were spontaneous, included examples which contained an amino acid replacing tryptophan in peptide No. 1. The amino acid substitutions were leu, ser, glN, glu, tyr, and lys or arg. In addition, some revertants were to polypeptide 1 with the original tryptophan present. H-12 was therefore a try → gap (amber) mutation, and the revertants were gap → leu, ser, glN, glu, try, tyr, and lys or arg. The codes for these amino acids are

UGe	try	CUd	leu	CAe	glN
UCd	ser	AAe	lys	GAe	glu
AGb	ser	CGd	arg	UAb	tyr
UUe	leu	AGe	arg		

Of these, the following, and no others, are all derivable from UAe by single-base changes: UGe, UCe, UUe, AAe, CAe, GAe, and UAb (try, ser, leu, lys, glN, glu, and tyr). Each of the amino acids that replaced tryptophan restored the enzymatic function of the alkaline phosphatase of *E. coli*. The report by Weigert and Garen therefore establishes the amber mutant as UAG or UAA, provides supporting evidence for the triplet sequences of six amino acids, describes seven new single-base mutations, and identifies all single amino-acid mutations from the chain-terminating triplets UAG and UAA with the possible exception of UGA. It was also stated that suppression is produced by the substitution of serine at the locus of the chain-terminating triplet in a phosphatase-negative mutant. If there is an "amber-sRNA" containing a triplet coding sequence of CUA, a possibility is that this sRNA might be changed so that it sometimes accepts serine, so that the suppressor mutation may be a change in the region of sRNA that recognizes the amino-acyl-activating enzyme complex or a change in the amino-acyl-activating enzyme itself. A second possibility is a change in the coding triplet of one of the serine sRNAs, so that it subsequently complements with UAG.

An investigation by Menninger (1965) showed that the amino-acyl-activating enzyme apparently was not altered in various amber suppressor mutants, for the su$^-$ enzymes completely unloaded su$^+$ sRNA that had been loaded with suppressor-specific amino acids by the su$^+$-activating enzymes. The reversibility of one enzyme by the other indicated that no change had taken place.

Capecchi and Gussin (1965) have provided additional evidence for participation of serine in the mechanism of suppression of an amber mutation. They found that a suppressor strain of *E. coli* permitted the formation of coat protein by an amber mutant of bacteriophage R17. The RNA isolated from the R17 mutant enabled protein synthesis to take place in a cell-free system derived from the suppressor strain but not in a system derived from an isogenic nonsuppressor strain. The difference was found to reside in the presence of a serine-accepting sRNA in the suppressor strain, this specific type of seryl-sRNA being absent from the nonsuppressor.

The effect of suppressor strain *Su-1* (Su_1^+) of *E. coli* was examined in chain-terminated mutants of another species, bacteriophage f2, by Notani et al. (1965). Again it was found that the suppressor inserted serine at an amino acid site that had mutated from glutamine to a chain-terminating triplet. The coat protein of the wild type contained a chymotryptic peptide thr.glN.phe which was replaced in the suppressed mutant su-3 by thr.ser.phe.

Weigert and Garen (1965b), in further studies with alkaline phosphatase, found that chain-terminated mutants could originate from

124   Mutations

changes in either a tryptophan or a glutamine locus, and in each case the suppressor strain *Su-1* would insert serine at the affected locus.

The various findings with respect to changes and suppression in the amber locus are summarized in Table 4-4. The conclusion by Brenner

Table 4-4. Relationships among the Amber Triplet UAG, Its Precursors, Its Reversion, and Its Suppression as Reported by Various Investigators

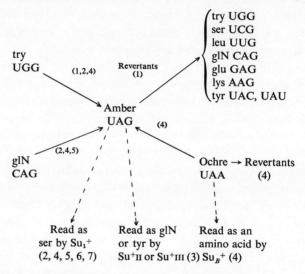

Sources:
1. Weigert and Garen (1965a).
2. Weigert and Garen (1965b).
3. Kaplan et al. (1965); Weigert et al. (1965).
4. Brenner, Stretton, and Kaplan (1965).
5. Notani et al. (1965).
6. Stretton and Brenner (1965).
7. Capecchi and Gustin (1965).

et al. (1965) that the amber triplet is UAG leads to the coding assignments in the table. A possible mechanism for certain suppressor mutations could be as follows:

A single-base change could take place in the coding triplet of a messenger RNA molecule, so that a G becomes changed to an A, which is then deaminated to I. This would produce an ambiguous triplet. A second sRNA would take over the function of supplying the amino acid to the messenger triplets formerly translated by the mutant sRNA (Crick, as cited by Kaplan et al., 1965), as shown in Table 4-5.

Table 4.5. Possible Suppressor Mechanisms

Postulated sRNA Coding Triplet and Pairing	Mutation to	Pairing with
CGA–ser, UCG	CUA	UAG
UGA–ser, UCA, UCG	UUA	UAA, UAG
CUG–glN, CAG	CUA	UAG
UUG–glN, CAA, CAG	UUA	UAA, UAG
AUA–tyr, UAU	CUA	UAG
AUA–tyr, UAU	UUA	UAA, UAG

The incorporation of serine by a UAI copolymer was studied by Takanami and Yan (1965) in two different cell-free systems obtained, respectively, from wild type (B) *E. coli* and from a suppressor mutant CR 63. About 40 percent more serine was incorporated by CR 63 than by B. Twelve other amino acids showed no such differences. All comparisons were on the basis of phe = 100.

Mutations and the Code

In 1957 (Ingram, 1957), a new doorway had opened to the study of mutations. This was disclosed by the findings with sickle-cell anemia, the "molecular disease" of Pauling, which was found to be caused by the change of a single-amino-acid link in a polypeptide chain containing 146 units. It was known that the sickle-cell trait was inherited as a simple nondominant allele. The discovery showed that the smallest phenotypic unit was a single amino acid. The inference was strong that a single-base change had given rise to a change in the meaning of a single unit in the code. Within the next eight years, more than 100 examples of identified mutational single-amino-acid changes at various loci in proteins were described in publications from various laboratories. This information was of great value in solving the coding problem, coming as it did before the trinucleotide-ribosome-sRNA binding technique that was rapidly exploited by Nirenberg's group (Leder et al., 1965; Trupin et al., 1965; Leder and Nirenberg, 1964; and Nirenberg et al., 1965); the studies of block copolymers by Nishimura et al. (1965), Thach, Sundararajan, and Doty (1965), and Thach and Sundararajan (1965); and the findings of Brenner et al. (1965) and Weigert and Garen (1965) with amber mutants of T4 bacteriophage.

The end of the search came abruptly in April, 1965, when the various findings fell into place in a coherent pattern, as described in Chapter 2 and Tables 2-11 and 2-13, which provide assignments for all 64 triplet permutations of A, C, G, and U. The single-amino-acid mutations corresponded to single-base changes in the triplets, with the sole exception of a spontaneous change, asN/arg, reported once in TMV which was subsequently shown to be asN/lys (Wittmann-Liebold and Wittmann, 1966). The universality of the code extended into all species of organisms that were examined.

It is probable that major changes in evolution leading toward greater complexity are brought about by duplications of genes occurring during meiosis. Simultaneously there is a steady series of single-base changes occurring throughout the genetic material. Any one of the four bases may become replaced by another, and this change may take place in either strand of DNA. The new base will be paired by its complement when the DNA molecule is next replicated. Whether or not the intrusion is tolerated will depend upon the effect, if any, that the change produces in phenotypic characteristics. Changes that produce no phenotypic effect may occur and revert imperceptibly. If the intrusion occurs in gametes, and if it is perpetuated, it may become established in the genetic material until, perhaps after a long time, another change occurs in the same pair of bases which may either restore the original pair or produce a new substitution.

Such changes can produce alterations in coding properties of triplet sequences in molecules of messenger RNA. It is also possible for such changes to alter the properties of the molecules of transfer RNA with resulting changes in the mechanism of protein synthesis.

An irregular (non-Mendelian) segregation of chromosomes at meiosis can directly alter allele frequencies in a population. This is termed *meiotic drive*, and a *segregation-distorter* gene has been described which causes unequal distribution of chromosome II at meiosis in *Drosophila* (Hiraizumi, Saudler, and Crow, 1960). Such events provide for the duplication of a whole chromosome. Duplication of part of a chromosome may also occur during meiosis, thus providing a mechanism for the duplication of a gene in one of the two haploid cells and its loss

Table 4-6. Separate Evolutionary Pathways in a Single Species; Comparison of Pairs of Proteins Produced by Gene Duplication

Protein Pair	Approximate Number of DNA Base Pairs Involved in Duplications [a]	Base Changes per 100 Base Pairs [b]
Myoglobin/αHb [c]	437	40
αHb/γHb [c]	437	25
γHb/βHb [c]	438	12
βHb/δHb [c]	438	2
Trypsinogen/chymotrypsinogen	747	25
α/β chains, insulin	69	25
Oxytocin/vasopressin	27	7
Ferredoxin, internal sequence, *Clostridium pasteurianium*	54	24

[a] Assuming all base deletions took place subsequent to duplication.
[b] Excluding deletions.
[c] Variable loci (Chapter 5) in human hemoglobins.

from the other. Studies of the primary structure of proteins have been completed for only a few proteins, but several clear examples of gene duplication have been revealed by these studies, including those of myoglobin and the hemoglobins, trypsinogen, the chymotrypsinogens, and elastase, oxytocin, and vasopressin, and the two chains of insulin (Chapter 6). The extent to which freedom from constraints has allowed the two members of each pair to evolve separately is indicated in Table 4-6. Partial gene duplication also exists in *Clostridium* ferredoxin (Fig. 6-5).

The present knowledge of the amino acid code enables a change to take place in the evaluation of point mutations. Formerly they were of value in deciphering the code; now the code will be used in the identification of the amino acids that are involved in mutations.

128 Mutations

Many evolutionary changes in proteins are the result of point mutations, and may be interpreted in terms of changes in the coding triplets. This will be discussed in Chapters 5 and 6.

Much information is available on mutational changes in proteins in which a single amino acid becomes substituted by another one. When all the coding triplets (Table 2-12) are compared, it is found that 82

Table 4-7. Amino Acid Interchanges That Can Occur as the Result of Single-Base Changes in the Coding Triplets

Amino Acid	Possible Interchanges[a]
ala	*asp* *glu* *gly* pro ser *thr* *val*
arg	cys *glN* *gly* *his* *ilu* *leu* *lys* met pro *ser* *thr* *try*
asN	*asp* his ilu *lys* ser thr tyr
asp	glu *gly* *his* tyr *val*
cys	*gly* phe ser *tyr* try
Gap	*glN* *glu* *leu* *lys* ser *try* *tyr*
glN	*glu* *his* leu *lys* pro
glu	*gly* lys val
gly	ser try *val*
his	leu pro *tyr*
ilu	leu lys *met* phe *ser* *thr* *val*
leu	*met* *phe* *pro* *ser* try val
lys	thr met
met	thr *val*
phe	ser *tyr* val
pro	ser thr
ser	thr try tyr

[a] The italicized examples have been reported to occur in mutations.

single-amino-acid changes, including those involving gaps (chain-terminating triplets), can be produced by single-base changes (Table 4-7). No less than 53, or 64 percent, of these changes have been described in mutations in the following proteins: human hemoglobin, TMV coat protein, A protein of tryptophan synthetase of *E. coli*, β-lactoglobulin of cow's milk, human cytochrome *c*, alkaline phosphatase of *E. coli*, malic dehydrogenase of *Neurospora*, and proteins of T4 and f2 bacteriophages of *E. coli*. This small group of proteins touches on a gamut of living organisms ranging from human beings to a virus that is parasitic on green plants. The findings are summarized in Table 4-8. Many of the 29 amino acid interchanges caused by single-base changes that have not yet been recorded in mutations are known to occur repeatedly in evolutionary changes in the molecules of hemoglobin (see Table 5-2).

Table 4-8. Single-Amino-Acid Mutations in Proteins

Amino Acid Change	Base Change	Protein and Location of Residue, If Known	Source
ala/asp	C/A	Hb α5	47
asp/ala	A/C	TMV 19 TS	11, 49, 50
ala/glu	C/A	Hb β70 or 76, δ22	1, 45
glu/ala	A/C	TS, Hb β43	2, 3, 4
ala/val	C/U	Lgb, AP	5, 6
val/ala	U/C	TS, TMV 19	2, 50
arg/gly	A/G, C/G	TMV 46, 61, 122, 134	7
gly/arg	G/A, G/C	TS, Hb δ16	2, 45
arg/ilu	G/U	TS	8
arg/lys	G/A	TMV 46	9
arg/ser	A/C, U	TS	2
arg/thr	G/C	TS	22
arg/try	C/U	MD	10
asN/asp	A/G	TMV 126	11
asp/asN	G/A	Hb β79	12
asp/val	A/U	TMV 19	51
val/asp	U/A	TMV 19	51
asN/lys	C, U/A, G	Hb α68, TMV 140	11, 13
lys/asN	A, G/C, U	Hb β61	14
asN/ser	A/G	TMV 25, 33, 73, 126	9, 11
asp/gly	A/G	TMV 66, Lgb, Hb α47	5, 11, 15
gly/asp	G/A	Hb α15, α22, α57, β16, TS	2, 16, 17, 18, 19
Gap/gap	A/G	T4	20
Gap/glN	U/C	AP, T4	20, 21
glN/gap	C/U	T4, f2	20, 22
Gap/glu	U/G	AP	21
Gap/leu	A/U	AP	21
Gap/lys	U/A	AP	21
Gap/ser	A/C	AP	21
Gap/try	A/G	AP, T4	20, 21
try/gap	G/A	AP, T4	20, 21
Gap/tyr	G/C, U	AP, T4	21, 23
glN/arg	A/G	Hb α54, TMV 99	9, 24
glu/glN	G/C	Hb α30, β121	25, 26
glN/glu	C/G	Hb α54	27

Mutations

Table 4-8. (continued)

Amino Acid Change	Base Change	Protein and Location of Residue, If Known	Source
glu/gly	A/G	Hb β7, TMV 97	11, 28
gly/glu	G/A	TS	2
glu/lys	G/A	Hb α116, β6, β7, β26, β121, γ5 or 6	28, 32, 33
glu/val	A/U	Hb β6, TMV 22* (?), TS	2, 11, 34, 35
val/glu	U/A	Hb β67	18
gly/ala	G/C	TMV	36
gly/cys	G/U	TS	2
gly/val	G/U	TS	2
his/arg	A/G	Hb β63, δ2	37, 45
his/asp	C/G	Hb β143	19
his/glN	C, U/A, G	Lgb	38
his/tyr	C/U	Hb α58, α87, β63	39, 40, 41
ilu/mct	A, C, U/G	TMV	11
ilu/ser	U/G	TS	49
ilu/thr	U/C	TMV 21, 129	7, 9, 11
thr/ilu	C/U	TMV 5, 59, 136, 153, TS	2, 11, 51
ilu/val	A/G	TMV 21, 24, 125, 129	9, 11
leu/arg	U/G	TS	2
leu/phe	C/U	TMV 10	11, 51
phe/leu	U/C	TMV 10	51
lys/glN	A/C	Hb β132	46
lys/glu	A/G	Hb α18, β95, β61	42, 43, 45, 51
met/leu	A/U	Cyt.c	44
pro/leu	C/U	TMV 20, 156	11, 36
pro/ser	C/U	TMV 63	11
pro/thr	C/A	TMV 20	9
ser/gly	A/G	TMV 65	9
ser/leu	C/U	TMV 15, 55, TS	2, 11, 50
ser/phe	C/U	TMV 138, 148	7, 9, 11
thr/ala	A/G	TMV 28, 81	9, 50
thr/met	C/U	TMV 107	11
thr/lys	C/A	Hb β87	48
tyr/cys	A/G	TS	2
tyr/phe	A/U	TMV	35
val/met	G/A	TMV 11	9

The 53 known interchanges that occur in mutations are in Table 4-8 and are shown diagrammatically in Fig. 4-4. All except five are due to base substitutions in one of the first two positions in the triplets, undoubtedly because changes in the third base in many cases do not change the coding properties and so are "silent." Such changes may, however, change the subsequent effect of a single-base change in one of the first two bases; for example, a change in an ala triplet from GCA to GCC could affect a subsequent change of C to A in the second base so that this change would be to GAC, asp, instead of GAA, glu.

Footnotes and Sources for Table 4-8:

* Formerly glN/val.
AP = alkaline phosphatase, *E. coli*; Lgb = bovine β-lactoglobulin; Cyt *c* = human cytochrome *c*; TS = tryptophan synthetase, *E. coli;* TMV = tobacco mosaic virus coat protein; Hb = hemoglobin; MD = malate dehydrogenase (*Neurospora*); T4 = *E. coli* bacteriophage T4; f2 = *E. coli* bacteriophage f2.

1. Huehns and Shooter (1965).
2. Yanofsky (1964).
3. Yanofsky (1963*a*).
4. Bowman et al. (1964).
5. Gordon et al. (1961).
6. Garen et al. (1965).
7. Tsugita (1962).
8. Yanofsky (1965).
9. Funatsu and Fraenkel-Conrat (1964).
10. Munkres and Richards (1965).
11. Wittmann and Wittmann-Liebold (1963).
12. Lehmann et al. (1964).
13. Baglioni and Ingram (1961).
14. Shibata et al. (1964).
15. Baglioni (1965).
16. Baglioni (1962*b*).
17. Gottlieb et al. (1963).
18. Chernoff and Perrillie (1964).
19. Shibata et al. (1963).
20. Brenner et al. (1965).
21. Weigert and Garen (1965).
22. Notani et al. (1965).
23. Stretton and Brenner (1965).
24. Hanadu and Rucknagel (1965).
25. Swenson et al. (1962).
26. Baglioni (1962*a*)
27. Lisker et al. (1963).
28. Baglioni and Lehmann (1962).
29. Hunt and Ingram (1959*a*).
30. Hill and Schwartz (1959).
31. Hunt and Ingram (1959*b*).
32. Pierre et al. (1963).
33. Schneider and Jones (1965).
34. Ingram (1957).
35. Tsugita (1962*b*).
36. Tsugita and Fraenkel-Conrat (1962).
37. Muller and Kingma (1961).
38. Kalan et al. (1964).
39. Gerald and Efron (1961).
40. Jones et al. (1964).
41. Ingram (1962, p. 179).
42. Beale and Lehmann (1965).
43. Krause (1965).
44. Matsubara and Smith (1963).
45. Jones et al. (1965, 1966).
46. Allan et al. (1965).
47. Crookston et al. (1965).
48. Watson-Williams et al. (1965).
49. Yanofsky et al. (1966).
50. Wittmann-Liebold and Wittmann (1965).
51. Wittmann-Liebold et al. (1965).

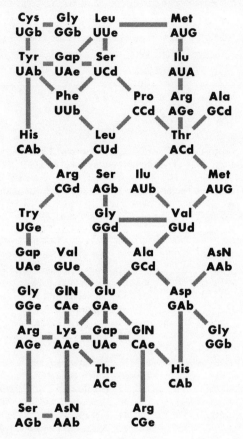

Fig. 4-4. The single-amino-acid mutations in proteins and their coding relationships (b = U, C; d = A, C, U, G; e = A, G).

The question of the occurrence of transversions is now settled by knowledge of the amino acid code. Twenty-five of the 46 known single-amino-acid mutations (excluding the "gap" mutations) result from transitions, including all possible transitional interchanges except arg/cys. These 25 interchanges are, however, weighted in favor of transitions by the fact that 18 of them are derived solely from nitrous-acid-induced mutations. The remaining 21 interchanges result from transversional (purine-pyrimidine) interchanges in the coding triplets. Numerous transversions have occurred in evolution. The hemoglobins and myoglobin genes in Table 5-2 show that at least 535 transversions have occurred in a total of 465 triplet sites, as compared with 285 perceptible transitional changes.

The mutations produced by HNO_2 in TMV are in most cases attributable to deaminative changes (A/H and C/U) in TMV RNA. In some cases, however, the changes correspond to G/A, U/C, or transversions. Evidently, the action of nitrous acid is not confined to deamination but can produce other types of changes in the strands of RNA, perhaps by damage and subsequent repair during replication. The report by Tessman et al. (1964) supports such a conclusion.

Mutations in Tryptophan Synthetase

A series of investigations by Yanofsky and his associates (1964) on the A protein of tryptophan synthetase have contributed to knowledge of the mutations occurring at a locus for a single amino acid.

The wild type of *E. coli* can grow in minimal culture media without amino acids. It can synthesize tryptophan and other amino acids from simple precursors. The final steps in the synthesis of tryptophan are carried out by the enzyme tryptophan synthetase, which has two protein components, A and B. These function in the following three reactions:

Indole + L-serine → L-tryptophan + H_2O
Indoleglycerol phosphate + L-serine
 → L-tryptophan + glyceraldehyde-3-phosphate
Indoleglycerol phosphate + H_2O
 ⇌ indole + glyceraldehyde-3-phosphate

The polypeptide sequence of the A protein contains 280 amino acids.

Certain mutations of *E. coli* caused by ultraviolet light and other effects make the A protein ineffective in the tryptophan synthetase reaction, so that the organism becomes dependent on an external source of tryptophan. It is possible to culture such mutants in media containing tryptophan. Defective forms of A protein in highly purified or crystalline forms may be obtained by extracting the bacteria in such cultures. These have been analyzed by Yanofsky et al. (1964), and in some cases the defect was shown to be due to the substitution of a single amino acid at a locus which is obviously critical in terms of enzymatic function.

The defective mutants can revert spontaneously to give rise to strains which are able to grow on a minimal medium just as in the case of the wild type. These new strains are of various types, some partially revertant and some fully revertant, as shown by growth characteristics and biochemical comparisons with the original wild strain. One critical locus that was studied intensively is at position 8 in chymotryptic peptide 2, which is occupied by glycine in A protein of the wild type. In two defective mutants (A-23 and A-46) the locus was found to be occupied by arginine and glutamic acid, respectively. These changes can be written as GGd/CGd or GGe/AGe (gly/arg) and GGe/GAe (gly/glu). The observation was made by Yanofsky and Helinski (1962) that if A23 and A46 mutants were crossed by a transduction procedure, a small proportion (1 in 10^6) of the bacteria resulting from the cross were able to grow on the medium without tryptophan and were therefore wild type. These contained A protein with glycine at the locus under examination, from which it could be deduced that crossing over of the genetic DNA had taken place within the coding triplet, so that one of the two offspring of the cross contained glycine. Since two G's are needed in the glycine code, the new triplet thus formed contained one G from A23 and one from A46. Thus, the gly/arg and gly/glu mutations had involved different G's in the glycine code, and the cross could be represented as

$$\begin{array}{cc} \text{arg AGe} & \text{GGe gly} \\ \times & \rightarrow \quad + \\ \text{glu GAe} & \text{AAe lys} \end{array} \quad (e = A, G)$$

or

$$\begin{array}{cc} \text{arg CGd} & \text{GGd gly} \\ \times & \rightarrow \quad + \\ \text{glu GAe} & \text{CAe glN} \end{array} \quad (d = A, G, C, U)$$

Further studies in other mutants showed that yet other amino acids could occupy the same locus; A23 (arg) produced mutants in which arg was replaced by ser, ilu, and thr, while A46 (glu) could give rise to mutants in which the glu was replaced by ala and val. Some of these were enzymatically active in the tryptophan synthetase reaction.

The activity varied so that the revertants were placed in two classes: fully revertant strains which had the same growth rate as the wild type

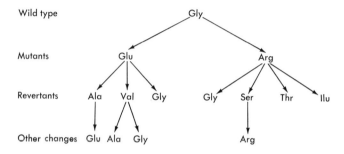

Fig. 4-5

and were resistant to inhibition by low levels of 5-methyl tryptophan, and partially revertant strains that grew more slowly than the wild type and were sensitive to 5-methyl tryptophan. Presumably, these two classes reflected differences in the secondary and tertiary structure of the A protein produced by the amino acid change. Substitution of glycine by alanine gave a fully active strain but by val a partially active strain; this finding may be compared with the fact that alanine resembles glycine more closely than does valine. Spontaneous reversion could take place to the "progenitor" amino acid in some of these, as shown in Fig. 4-5 (Yanofsky, 1965). The numbers by the arrows show how often each change was found. The arginine code that corresponds to the changes is AGA, and the mutations may be written as shown in Fig. 4-6. Transduction studies between selected pairs of the mutants were used to produce new variants which included wild type in some

136 Mutations

cases. For example, transduction between arg and val mutants was found to produce gly and ser. This finding cannot be explained in terms of crossing over between arg and val triplets, and some form of spontaneous reversion presumably must have taken place during the experiment.

Yanofsky (1963) proposed that certain restrictions be placed on the nature of the base substitutions in these changes on the basis of their

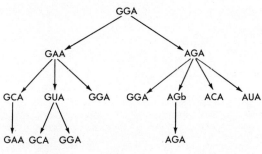

Fig. 4-6

production by the chemical mutagens 2-aminopurine (AP), 5-bromodeoxyuridine (BU), and ethyl methane sulfonate (EMS). It was found that with A46 the glu/gly change was the most frequent change with AP and was more frequent than were the spontaneously occurring changes, suggesting that glu/gly represented a transitional (pu/pu or py/py) change at the DNA level. This would be in agreement with a GAA/GGA change for this revertant. It was also found that AP favored the arg/gly change with A23. This indicated that this represents AGe/GGe rather than CGd/GGd. On the other hand, glu/val or ala (GAe/GUe or GCe) and arg/thr or ser (AGe/ACe or AGb) were not favored by AP, from which it was inferred that these were transversional (pu/py or py/pu) changes, which is supported by the coding assignments in Fig. 4-6. A series of replacements at another locus in tryptophan synthetase A protein was described by Guest and Yanofsky

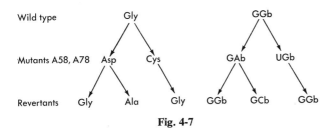

Fig. 4-7

(1965) as in Fig. 4-7. Recombination between A58 and A78 yielded wild type (gly); GAb × UGb → GGb +? (tyrosine for "?" would be predicted).

Mutations in the Protein Coat of Tobacco Mosaic Virus

Tobacco mosaic virus has served as a molecular model of a living system from the time since it was first crystallized by Stanley (1935). The virus consists of a long single molecule of RNA containing a sequence of about 6,000 bases and surrounded by a tubular closely packed spiral, or "coat," of protein, consisting of about 2,100 repeating units. The coat protein apparently has protective and possibly other functions. The particle of the virus consists of a rod with a radius of 150 to 180 A and a length of 3,000 A with a molecular weight of 40 million, 95 percent of which is due to protein. Each protein unit therefore has a molecular weight of about 18,000. Since the genetic code consists of triplets, 474 nucleotides would be sufficient to code for the coat protein of TMV, since each unit contains 158 amino acids. The entire amino acid sequence of the coat protein is known (Anderer, 1962; Knight, 1963; Wittmann-Liebold et al., 1965).

When TMV enters and infects a cell, it is assumed that the virus sheds its protein coat and the RNA strand proceeds to replicate. This process in RNA viruses has been found to take place following the formation of a double-stranded *replicative form* consisting of the original single-stranded virus joined to its complement (Montagnier and Sanders,

1963; Weissmann and Borst, 1963). The replicative form combines with an enzyme, RNA synthetase, and proceeds to make quantities of new single strands, each of which is identical with the original parental type. The DNA of the host is not involved in these procedures.

The single strands of virus RNA parasitize the protein synthesis equipment of the whole cell and proceed to make proteins for the benefit of the virus. It is presumed that this procedure involves combining with the host cell ribosomes and utilizing the sRNA and transfer enzymes of the host. Obviously the virus, to be successful, must use the same code as the host, or it could not utilize the host-cell sRNA. We do not know what proteins are made by the virus sRNA, except that one of them is the coat protein. The coding requirement for this would correspond to about 7.4 percent of the base sequence of TMV RNA, and the remaining 92.6 percent is available for coding other proteins, including enzymes, one of which is perhaps a lysozyme-like enzyme that enables the completed virus to dissolve its way out of the cell. We may guess that the TMV RNA surrounds itself with its protein coat either as it detaches itself from a polysome or immediately thereafter.

The infectivity of the virus is enhanced by the presence of the protein coat, but the RNA molecule retains some infective power when the coat is removed by chemical means, such as treatment with dodecyl sulfate. The infective process is due to the TMV RNA functioning as messenger RNA within the host cell. As a result, the proteins necessary for the replication of the virus, including the coat protein subunit, are synthesized within the host cell.

TMV has been used extensively as a subject for chemically induced mutations. Treatment of TMV or of TMV RNA with mutagenic agents, especially nitrous acid, inactivates most of the particles in the virus preparation but a small percentage produce infections. It was found by Schuster and Schramm (1958) that treatment with nitrous acid produced deaminative changes in TMV RNA so that cytosine was converted to uracil, adenine to hypoxanthine, and guanine to xanthine. Simultaneously, mutations were produced (Mundry and Gierer, 1958), so that the lesions produced by surviving particles were in some cases different in appearance from lesions produced by the untreated TMV RNA. Treatment of the intact virus with HNO_2 also produced mutants. Changes in the amino acids of the coat protein were found in some mutants produced by either method. This immediately made it probable that the amino acid changes were the direct result of the effect of nitrous acid on single bases in TMV RNA. The base changes of cytosine to uracil and adenine to hypoxanthine would alter the coding

Table 4-9. Distribution of Amino Acid Changes in HNO_2 and Spontaneous Mutants of TMV Coat Protein

Amino Acid Exchanges in Protein Coat Subunit	Number of Mutants	
	HNO_2	Spontaneous
0	88	8
1	31	5
2	6	2
3	0	1
>3	0	0

Source: Wittmann and Wittmann-Liebold (1963).

functions of the RNA triplets in many cases, since hypoxanthine behaves like guanine in Watson-Crick pairing.

The case of guanine to xanthine is more difficult to assess. Xanthine apparently does not function in coding, so that a change of guanine to xanthine might completely abolish the coding function of the triplet, thus arresting the synthesis of the protein with lethal results. It also seems possible that guanine is not always converted to xanthine by nitrous acid, so that perhaps other groups, as yet unidentified and with unknown coding functions, might replace guanine in the RNA strand. Wittmann and Wittmann-Liebold (1963) give the figures in Table 4-9 for the distribution of amino acid changes in HNO_2 and spontaneous mutants.

In no case were two adjoining amino acids found to be altered, an observation which demonstrated that the code is not overlapping; i.e., a coding triplet directs only one amino acid. Evidently, most HNO_2 mutants are altered elsewhere than in the coat protein, which is predictable in view of the consideration that only about one-twelfth of the

Mutations

RNA sequence is needed to code for the coat protein subunit sequence of 158 amino acids.

Most of the amino acid changes produced by HNO_2 in TMV correspond to deaminative base changes in the RNA, C/U, or A/hypoxanthine (equivalent to A/G in complementarity). Table 4-10 is based

Table 4-10

Amino Acid Replacement	Times Observed	Change in Coding Triplet
thr/ilu	9	C/U
ser/phe	4	C/U
pro/leu	3	C/U
pro/ser	3	C/U
ser/leu	2	C/U
asp/gly	2	A/G
thr/met	3	C/U
asN/ser	2	A/G
asp/ala	4	A/C
ilu/val	3	A/G
ilu/met	1	A/G
leu/phe	1	C/U
glu/val	2	A/U
glu/gly	1	A/G

Source: Wittmann and Wittmann-Liebold (1963).

on findings by Wittmann and Wittmann-Liebold (1963); 27 of the 40 observed changes correspond to deamination.

It is not possible to write asp/ala and glu/val as single-base deaminative changes, so that it is concluded that nitrous acid may produce other changes in RNA in addition to deaminations. Geiduschek (1961) has also reached this conclusion.

Chemically evoked mutants of TMV have also been studied extensively in the laboratory of Fraenkel-Conrat (Tsugita and Fraenkel-Conrat, 1960; Tsugita, 1962), and the findings were summarized by Funatsu and Fraenkel-Conrat (1964). Bromination was carried out by N-bromosuccinimide (NBSI), deamination by nitrous acid (HNO_2), and methylation by dimethylsulfate (DMS). The results are given in Table 4-10.

Again, the majority of the changes induced by nitrous acid correspond to deaminative single-base changes in the coding triplets. The exceptions are arg/lys and val/met.

A consideration of the possible effects of nitrous acid upon mutations led Gierer (1961) and Wittmann (1962) to arrange the 64 possible triplet permutations of the four RNA bases, A, G, C and U, into eight

Table 4-11

Amino Acid Replacement	Chemical Mutagen	Location of Residue	Inferred Change in Coding Triplet
arg/gly	HNO_2, Me	46	A/G
arg/lys	HNO_2	46	G/A
asN/ser	HNO_2, NBSI	25, 33, 126	A/G
glN/arg	HNO_2	99	A/G
glu/gly	HNO_2	97	A/G
ilu/thr	NBSI	21	U/C
ilu/val	HNO_2	21, 129	A/G
pro/leu	HNO_2, NBSI	20, 156	C/U
pro/thr	Spontaneous	20	C/A
ser/gly	NBSI	65	A/G
ser/phe	HNO_2, Me	138, 148	C/U
thr/ala	HNO_2, Me	81	A/G
val/met	HNO_2	11	G/A

Source: Funatsu and Fraenkel-Conrat (1964).

octets (Fig. 4-8). Each octet depicts 12 deaminative changes, each deamination involving a single-base transition. These give rise to a total of 26 different interchanges of amino acids, all of which are known to occur in one or both directions in mutations with the exception of arg/cys (Table 4-12). Those marked * have been detected in TMV mutants (Table 4-8) in either or both directions. The number of different single-base changes per amino acid interchange in the octets may be either one (e.g., thr/met), two (e.g., ser/phe) three (e.g., thr/ilu), or four (e.g., pro/leu). This may have a bearing on the frequency of the respective occurrence of the amino acid interchanges. In addition, 49 amino acid interchanges may occur as a result of transversional (pu/py and py/pu) single-base changes. Four amino acid interchanges, arg/gly, arg/try, ilu/met, and leu/phe, may occur because of either

142 Mutations

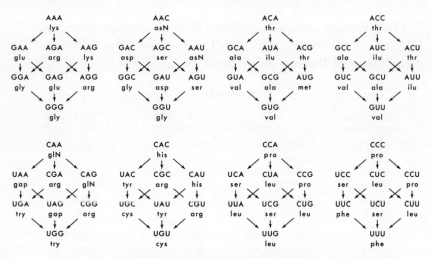

Fig. 4-8. The "octets," showing all possible deaminative single-base changes in the coding triplets.

transitions or transversions, as may be seen by examining Table 2-12. This occurrence of three of these four in TMV mutations caused by nitrous acid is known, and it is tempting to attribute them in this case to transitional changes.

Crossing over between synonymous coding triplets does not change their coding assignments except in the cases of arg, leu, and ser, each of which has two doublets in its codes. The following are the possible

Table 4-12

Octet: 1	2	3 and 4	5	6	7 and 8
* lys/arg	* asN/ser	* thr/ala	(glN/gap)	his/tyr	* pro/ser
lys/glu	* asN/asp	* thr/ilu	* glN/arg	his/arg	* pro/leu
* arg/gly	* ser/gly	* ilu/val	arg/try	arg/cys	* ser/leu
* glu/gly	* asp/gly	* ala/val	(gap/try)	tyr/cys	* leu/phe
		* thr/met	(gap/gap)		* ser/phe
		* ilu/met			
		* met/val			

cross-overs in these cases:

	arg	CGb →	AGb	ser			
×	arg	AGe	+ CGe	arg			
	ser	UCe →	AGe	arg,	or	ACe	thr
×	ser	AGb	+ UCb	ser		+ UGb	cys
	leu	UUe →	CUe	leu			
×	leu	CUb	+ UUb	phe			

It is conceivable that two genes that were identical except for two different ser codes at the same locus could cross over to produce either one or two new phenotypes as shown above. This point has been stressed by Sonneborn (1965). For two genes to have different ser codes at the same locus, however, it would be first necessary for one ser code to change into another through the code for a different amino acid, such as ser UCC → cys UGC or thr ACC → ser AGC. This two-stage sequence seems no more probable than any of the other 81 possible two-step changes from UCC to other triplets, so that the probability of a ser × ser crossover producing arg + ser or thr + cys seems vanishingly small. The possibility for transition from one code to another is greater in the cases of arg and leu, but here the cross-over (arg → ser or leu → phe) will have no more effect than a point mutation which can produce the same change. A mutation of one synonymous triplet to another, amber to ochre, was detected by Brenner et al (1965).

Mutations in the Lysozyme Gene of Bacteriophage T4

It was proposed by Brenner et al. (1961) that the mutagenic effect of acridine was due to the insertion or deletion of base pairs in DNA, corresponding to adding or subtracting single bases in messenger RNA. This would produce a shift in the reading frame of the messenger so that all triplets distal to the change would be out of step. A biologically useless protein would be produced unless the disturbance were close to the end of the polypeptide, in which case the change would affect only a few amino acids and perhaps would be unnoticed. Furthermore, if a deletion were made by a mutagen, and then an insertion were made a few bases further on, only a few amino acids would be changed between the two mutated sites, and the distal portion of the protein would be normal (Crick et al., 1961).

The technique was applied to acridine mutants of the lysozyme gene in T4 bacteriophage by Terzaghi et al. (1965). Two lysozyme-negative mutants, eJ44 and eJ42, were recombined to produce a new mutant, eJ44J42, which was lysozyme-positive, with about 50 per cent of the lysozyme activity of wild type. The lysozymes were isolated from the wild type and eJ44J42 mutant and digested with trypsin and chymotrypsin. Peptides 10 and 19 were found to differ as follows between the wild type and mutant:

Wild:
  Chymotrypsin   thr lys ser pro ser leu asN
  Trypsin   ser pro ser leu asN ala ala lys COOH
eJ44J42:
  Chymotrypsin   thr lys val his his leu
  Trypsin   val his his leu met ala ala lys COOH

Evidently, the sequence ser pro ser leu asN had been changed to val his his leu met. This corresponds to the deletion of A at the left of the section coding for the sequence and the insertion of G at the right, with only one possible solution in terms of the code as follows:

...A A e A G U C C A U C A C U U A A U G C d...
       lys   ser   pro   ser   leu   asN   ala

...A A e G U C C A U C A C U U A A U G G C d...
       lys   val   his   his   leu   met   ala

The results also demonstrate that the genetic message is translated from left to right in protein synthesis, in terms of the convention used in writing the abbreviated formula for RNA, which is in the direction of 3'- to 5'-phosphate diester linkages in the examples shown above. Peptide 19 ends in . . . ala ala lys COOH, terminating with the carboxyl group of lysine, so it is evident that this was the last amino acid to be incorporated in this peptide, and its synthesis took place by reading the mRNA from left to right.

This remarkable experiment emphasizes the achievements that biochemistry can bring when brought to bear on genetics. The flexibility of the third base in the triplet is well illustrated; the mutant sequence contains examples of A, C, U, and G in the third position.

The findings make it possible to write the actual formula for a piece of a gene, using the abbreviations explained in Table 2-1, as follows:

$$\left\{\begin{array}{l}\cdots pApGpTpCpCpApTpCpApCpTpTpApApTpGpCp\cdots \\ \|\|\|\|\|\|\|\|\|\|\|\|\|\|\|\|\|\|\|\|\|\| \\ \cdots pTpCpApGpGpTpApGpTpGpApApTpTpApCpGp\cdots\end{array}\right\}$$

Yanofsky et al. (1966) have described a "mutator gene" in *E. coli*. The gene favored the changes arg → ser, asp → ala, cys → gly, glu → ala, and ilu → ser in mutations of the A protein. All these changes correspond to transversions $\frac{A}{T} \to \frac{C}{G}$ in the DNA. It is conceivable that such a mutator gene could be produced by a mutation in DNA polymerase (Speyer, 1965).

[5]

Evolution and the Hemoglobins

> But somewhere, beyond Space and Time
> Is wetter water, slimier slime!
> And there (they trust) there swimmeth One
> Who swam ere rivers were begun,
> Immense, of fishy form and mind,
> Squamous, omnipotent, and kind.
>
> Rupert Brooke, "Heaven"

Hemoglobin is the most conspicuous substance in blood; about 14 to 16 percent of blood consists of hemoglobin, which is present exclusively in the red blood cells. The function of hemoglobin in transporting oxygen from the lungs to the tissues makes it a key compound in the chemistry of life, and in consequence it has long been studied by many biochemists. Myoglobin is a similar substance that is present in muscle. Hemoglobin was one of the first proteins to be crystallized. In 1849, K. E. Reichert described tetrahedral crystals of what is now known as hemoglobin in the fetal membranes of the uterus of a guinea pig which was examined six hours after death. He measured and recorded the angular inclination of the planes of the crystals. Shortly afterwards, Funke (1851) described methods for crystallizing the substance from the blood of various species and noted that the forms and stabilities of crystals from different species were not alike. The substance crystallized by Reichert and Funke was named *hemoglobin* by Hoppe-Seyler (1864), who showed that certain reagents would decompose it into an albuminous substance and hematin (Hoppe-Seyler, 1862).

An extensive comparison of the hemoglobins of various species by crystallographic procedures was made by Preyer (1871), who noted that most species yielded rhombic crystals. It was, however, E. T. Reichert (Fig. 5-1), a professor of physiology, and A. P. Brown, a professor of geology and mineralogy, at the University of Pennsylvania, who first perceived the taxonomic significance of the crystallography

Fig. 5-1. Professor Edward Tyson Reichert. (Courtesy of the University of Pennsylvania.)

of hemoglobin and compiled the information on the crystal structure (1909) and angles of the vertebrate hemoglobins on a taxonomic basis.

Reichert and Brown were drawn to this method of taxonomic investigation by the ease of crystallization of the hemoglobins and by the principle that *"substances that show differences in crystallographic structure are different chemical substances* (emphasis ours)." Their voluminous monograph records a remarkable achievement. They found

148 Evolution and the Hemoglobins

that the crystals of the various species of any genus belonged to the same crystallographic system and generally to the same crystallographic group; they had approximately the same axial ratios, or their ratios were in a simple relation with each other. An example of comparisons is shown in Table 5-1.

Table 5-1. Crystallographic Comparison of Reduced Hemoglobins of Species in *Felidae* Contrasted with a Few Other Species of Carnivora

Specific Name	Common Name	Axial Ratio $a : b : \dot{c}$
Felidae:		
Felis leo	Lion	0.9742:1:0.3707
Felis tigris	Tiger	0.9742:1:0.3839
Felis bengalensis	Leopard	0.9657:1:0.3667
Felis pardalis	Ocelot	0.9489:1:0.3931
Felis domestica	Domestic cat	0.9656:1:0.3939
Lynx canadensis	Lynx	0.9605:1:0.3944
Lynx rufus	Wildcat	0.9869:1:0.3914
Canidae:		
Canis familiaris	Dog	0.6745:1:0.2863
Vulpes fulvus	Red fox	0.6494:1:0.2894
Ursidae:		
Ursus americanus	Black bear	1.2239:1:1.1429
Otariidae:		
Phoca vitulina	Harbor seal	1.2131:1:1.1970

Source: Reichert and Brown (1909).

An increase in the divergence of crystallographic properties was found to be parallel to the taxonomic separation of various animals. Of much interest is the fact that a sample of blood labeled as that of a baboon was found upon examination of the hemoglobin crystals to be that of a cat, and a subsequent followup showed that mislabeling of the sample vial had occurred (Reichert and Brown, 1907).

More than one polymorphic crystalline form of the same reduced hemoglobin or oxyhemoglobin from a single species was noted in a number of cases by Reichert and Brown, presumably because of differences in inorganic salts. This complicated their findings, but their monograph remains as the earliest landmark in the history of molecular evolution.

Hemin crystals obtained from different species were always the same, so that the differences were due to the globin portion of the molecule.

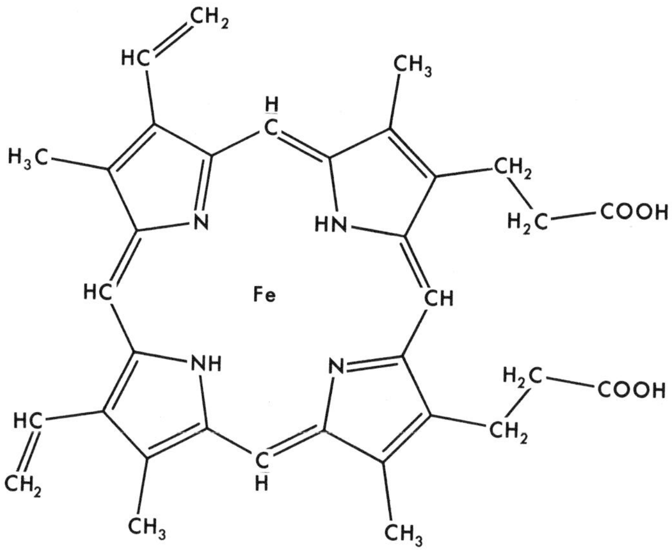

Fig. 5-2. Iron protoporphyrin IX, the prosthetic group of myoglobin and the hemoglobins.

It is now known that the differences are due to amino acid substitutions throughout the polypeptide chains of the globins. These substitutions are the result of single-base changes in the DNA strands of the hemoglobin genes. The hemoglobins constitute a chemical microcosm for the study of genetics, mutations, and evolution. The hemoglobins and myoglobins are conjugated proteins containing the prosthetic group *heme* (Fig. 5-2), which is protoporphyrin IX with iron. The protein portion of the molecule is *globin*. The iron content of hemoglobin was found by Engelhart in 1825 to be 0.35 percent, giving a value of about 16,000 for the minimal molecular weight of hemoglobin.

About 100 years later a molecular weight of about 64,000 to 68,000 was found for various vertebrate hemoglobins. An exception is that of the lamprey, a primitive chordate which contains a hemoglobin that resembles myoglobin in having a molecular weight of about 17,000. These and other observations show that the hemoglobins, except for lamprey hemoglobin, are tetramers consisting of four globin chains each combined with a heme group. Myoglobin and lamprey hemoglobin are monomeric forms of the globin-heme configuration. The tetrameric hemoglobins can be dissociated into two dimers and thence into monomeric units in solutions containing urea or guanidine or under conditions of low pH (Braunitzer et al., 1964). The globin portion of each tetramer typically contains two pairs of identical monomers. In the most common form of human hemoglobin, which is termed *adult hemoglobin A*, the molecule contains two α and two β chains, each of which is combined with a heme group. The abbreviation for the molecule is $\alpha_2^A \beta_2^A$, which means two human (A) α-hemoglobin polypeptide chains associated with two normal (A) β-hemoglobin polypeptide chains.

The four units are closely approximated in quaternary structure in the molecule by interpeptide salt linkages and various non-ionic interactions but not by covalent linkages. The tetrameric structure is regarded as an evolutionary advantage in that it brings the four heme groups into interaction and increases the efficiency of oxygen transport by leading to a stepwise association of oxygen with the molecule (Wyman, 1963).

The concept that each protein from each species of animal was a single chemical substance at the molecular level was implicit for the hemoglobins in the report by Reichert and Brown. It was again stated in 1952 by Sanger as a result of his studies of the amino acid sequence in insulin:

It has frequently been suggested that proteins may not be pure entities but may consist of mixtures of closely related substances with no absolute unique structure. The chemical results so far obtained suggest that this is not the case and that a protein is really a single chemical substance, each molecule of one protein being identical with every other molecule of the same pure protein. Thus it was possible to assign a unique structure to the phenylalanyl chains of insulin. Each position in the chain was occupied by only one amino acid and there was no evidence that any of them could be occupied by a different residue. Whether this is true for other proteins is not certain but it seems probable that it is. The N-terminal residue of several pure proteins have been determined . . . and this position is always found to be occupied by a single unique amino acid. These results would imply an absolute specificity for the mechanisms responsible for protein synthesis and this should be taken into account when considering such mechanisms.

The development of biochemical information in recent years has made the acceptance of this conclusion appear obvious, but seven years later Vaughan and Steinberg (1959) quoted two contrasting viewpoints: one had stated in 1953 that although certain groups in the protein must be correctly spaced and oriented for physiological function, "the identity of the remaining residues . . . may depend upon the environmental conditions of the cell at the moment of synthesis," and another had contended in 1954 that "a unique structure and configuration of proteins must be interpreted always in a statistical, not a definite sense."

The ease with which hemoglobin and myoglobin may be obtained in large quantities and crystallized made them prime targets for the analytical studies of amino acid sequences by Braunitzer et al. (1961), Edmundson (1965), Hill, Buettner-Janusch, and Buettner-Janusch, (1963), Konigsberg, Goldstein, and Hill, (1963), and Schroeder et al. (1961); and for the remarkable X-ray crystallographic investigations by Kendrew (1962) and Perutz (1962), which led to the elucidation of the complete primary, secondary, and tertiary structures of the hemoglobins and myoglobin and the quaternary structure of hemoglobin A. These researches have furnished the subject matter for a number of excellent reviews and monographs (Baglioni, 1963; Braunitzer, 1965; Braunitzer et al., 1964; Ingram, 1963; Kendrew, 1962; Perutz, 1965; Perutz et al., 1965; Zuckerkandl, 1964; Zuckerkandl and Pauling, 1962), and it would be inappropriate to recapitulate much of this material. Most of the attention in this chapter will be directed to the possible relationships of the amino acid sequences in the globins to the ordering of bases in the corresponding genes.

The molecule of sperm whale myoglobin is described by Kendrew (1962) as being arranged as a compact, rather triangular, flattened prism with dimensions about 45 × 35 × 25 A. Kendrew's description of the architecture of the molecule may be summarized as follows: The heme group is disposed almost normally to the surface of the molecule, with one of its edges at the surface and the rest buried deeply within. The molecule that is arranged into the prism is made of a single polypeptide chain of about 152 amino acid residues. Such chains usually contain regions in which the polypeptide sequence is wound into α helices, a configuration described by Pauling and Corey (1961). In the myoglobin molecule, 118 amino acid residues are in eight helical segments, each containing from 6 to 25 residues. The chain terminates with a nonhelical tail of 5 residues at the carboxyl end of the chain. In contrast to certain other proteins, myoglobin is not held together in tertiary structure by covalent S-S cystine cross-links. The helical content of the molecule is unusually high. The characteristic tertiary structure of the molecule depends upon the arrangement of the polar and nonpolar side chains of the various amino acids in polypeptide sequence; the polar groups are on the surface of the molecule, and the interior of the molecule is made up of nonpolar groups. Bound water groups are attached to all polar groups on the surface, but there are fewer than five water molecules inside the structure, and there are no channels through it (Kendrew, 1962), although water is present in the central cavity (Perutz et al., 1965). A representation of the tertiary structure is in Fig. 5-3.

It might well be anticipated that this unusual molecule would be unique; a structure whose functions are so highly specialized that its atomic configuration would not be subject to variation. Nothing could be further from the facts; many variations of the globin molecule, most of them present in hemoglobins, occur in different animals and have been analyzed and catalogued. Each of them has the same general prismatic tertiary structure that is described above for myoglobin; each contains the same number of similar helical regions linked by turns. Most remarkable of all, the various hemoglobins each consist of four myoglobinlike units, two α chains and two β chains, neatly packed into an almost spherical molecule, measuring 64 × 55 × 50 A and with the four heme groups in separate locations on its surface. In view of the interlocking arrangement of the four units, one would be inclined to conclude that the molecule had evolved as a tetramer and that myoglobin represented a later form which had dissociated from

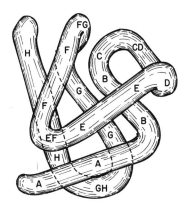

Fig. 5-3. Diagram of the configuration of the myoglobin molecule. The letters indicate helical regions. (See title of Table 5-2.) Adapted from Dickerson (1964).

the original quartet. Actually, the reverse is the case; the monomeric form of the molecule is more primitive than the tetramer, and the monomer exists as a presumably archetypal form in the lamprey. It appears that during the course of evolution, by what seems to have been a fortunate series of mutational events, modifications of the primitive molecule appeared that were able to associate into tetrameric groups, thereby increasing their effectiveness in oxygen transport. Furthermore, it is not necessary that the tetramer be composed of two pairs of different units; β_4 and γ_4 hemoglobins are found in genetically abnormal human beings.

The conclusion is drawn that the genes responsible for the production of the globin portion of the myoglobins and hemoglobins are all derived from a common archetypal piece of DNA, probably containing about 486 base pairs. This conclusion is based on the following evidence:

1. The polypeptide chains of the globins are all of about the same length and arrange themselves into similar secondary and tertiary structures. Each that has been examined contains two histidine residues, which are 29 residues apart and which bind the prosthetic heme group to the polypeptide chain.

2. Each globin contains about 9 loci in which the same amino acid is present in all globins from all species so far examined. These loci occur at similar positions. Starting at the NH_2 end of the chains, the first such locus is the glycine residue about 26 residues from the end. Thereafter, identical amino acids occur at varying intervals when the primary structures are aligned (Table 5-2). The existence of a few gaps in the chains is revealed by this alignment procedure. These gaps are considered to be due to the deletions of DNA base pairs in multiples of three during evolution; this hypothesis receives support from the fact that the gap which occurs between positions 46 and 47 in a number of mammalian α chains is occupied by an alanine residue in carp α hemoglobin chain (Braunitzer, 1965). Deletions of xy consecutive base pairs, unless x or $y = 3$, would displace the reading frame of the messenger so that a functional protein would not be produced; deletion of base pairs in multiples of 3 removes portions of the protein chain while leaving the amino acid sequence to the right of the deletion intact.

3. As the evolutionary tree is traced backward, increasing differences are found between the homologous globin chains of different species. Only one or two differences exist between the α hemoglobin chain of human beings and those of the gorilla and the orangutan, but there are 17 differences between the α hemoglobin chains of human beings and of horses and an equal number between the myoglobins of the human being and the sperm whale (Hill, 1963).

4. The differences in terms of amino acid substitutions are even

greater between the α and β hemoglobin chains of human beings, and between human myoglobin and the chains of human hemoglobins, than between the chains of human α hemoglobin and horse α hemoglobin. Furthermore, the difference between myoglobin and the α chain is about the same as the difference between myoglobin and either the β or γ chains, although the α, β, and γ chains all differ markedly from each other. These relationships are expressed in the "family tree" diagram (Fig. 5-4) of Ingram (1963, p. 143).

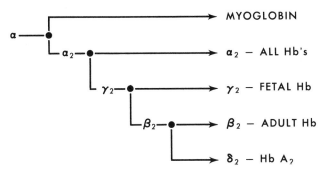

Fig. 5-4. Evolution of the hemoglobin chains. The point in time of a gene duplication is indicated by a solid black circle.

5. A number of mutational variants have been discovered in contemporary human hemoglobins, each characterized by a change in a single amino acid residue in either the α or β chains. The possessors of these variant hemoglobins can transmit them as simple nondominant alleles. Each single-amino-acid mutation can be translated into terms of a single-base change in the genetic code (Table 5-6). In five cases, two different mutations in hemoglobin chains occur at the same site, each corresponding to a single-base change. These observations may be regarded as an example of evolution in progress; but in this instance there is no reason to conclude that any of these mutations will be adopted into the human genome. They may be a part of the ebb and flow of changes that are incidental to the presence of mutational

Table 5-2. Amino Acid Sequences of Human Myoglobin and Alpha, Beta, and Gamma Hemoglobin, Together with Amino Acid Substitutions in the Corresponding Globins of Human Delta Hb, of Mutations, and of Other Species

Residue No.:	1	2				6					10				15		
Protein																	
Myo	NH₂-gly	leu	ser	glu	gly	glu	try	glN	leu	val	leu	his	val	try	ala	val	glu
α-Hb	NH₂-val	leu	ser	pro	ala	asp	lys	thr	asN	val	lys	ala	ala	try	gly	val	gly
β-Hb	NH₂-val his	leu	thr	pro	glu	glu	lys	ser	ala	val	thr	ala	leu	try	gly	val	
γ-Hb	NH₂-gly his	phe	thr	glu	glu	asp	lys	ala	thr	ilu	thr	ala	leu	try	gly	val	
Others	AcNH-val thr	ser		ala	lys	gly			his	leu	asN	ser	glu	phe	ser	leu	
	thr glN			gly	val	lys			gly		gly		arg		thr	ilu	
	leu			ser	asp				asp		asp				asp		
	tyr								lys		arg				lys		
	arg														arg		
Myo	GGd	CTb	TCd	GAe	GGe	GAe	TGe	CAe	CTb	GTb	TTe	CAb	GTe	TGe	GCb	AAe	GTb GAe
α-Hb	GTd	CTb	TCd	CCe	GCe	GAb	AAe	ACe	AAb	GTb	AAe	GCb	GCe	TGe	GGb	AAe	GTb GGe
β-Hb	GTd CAb	CTb	ACd	CCe	GAe	GAe	AAe	TCe	GCb	GTb	ACe	GCb	CTe	TGe	GGb	AAe	GTb
γ-Hb	GGd CAb	TTb	ACd	GAe	GAe	GAb	AAe	GCe	ACb	ATb	ACe	TCb	CTe	TGe	GGb	AAe	GTb
Others	ACd ACb			GCe	AAe	GGd	GAb		CAb	CTb	AAb		GAe	TTb	AGb	GAe	CTb
	CAe			GGe	GTe	AAe			GGb		GGb		CGe		ACb	GAb	ATb
	CTd			TCe	GAb				GAb		GAb				GAb		
	TAb								AAe		AGe				AAe		
	CGb														CGb		

Residue No.: 20, *26, 30, *37

Protein																	
Myo	pro(19)	asp	val	ala	gly	his	asp	ilu	leu	ilu(30)	arg	leu	phe	lys	gly	his	pro(37)
α-Hb	ala	his	ala(20)	gly	glu	tyr	ala	ala	leu	glu	arg	met	phe	leu	ser	phe	pro
β-Hb			asN	val	glu	val	gly	ala	leu	gly	arg	leu	leu	val	val	tyr	pro
γ-Hb			asN	val	asp	ala	gly	thr	leu	gly	arg	leu	leu	val	val	tyr	pro
Others	gly			glu	lys	thr	glN	ser	ilu	glN	lys	phe	thr				
	ser						ser			lys							

Protein																	
Myo	CCb	GAb	GTe	GCd	GGd	CAb	GGd	GAb	ATb	CTb	ATA	AGe	TTG	AAe	GGb	CAb	CCd
α-Hb	GCb	CAb	GCe	GGd	GAe	TAb	GGd	GCb	GCb	CTb	GAA	AGe	ATG	TTb	AGb	TTb	CCb
β-Hb			AAb	GTe	GAb	GTb	GGd	GAe	GCb	CTb	GGA	AGe	TTG	CTb	GTb	TAb	CCd
γ-Hb			AAb	GTe	GAe	GCb	GGd	GAe	ACb	CTb	GGA	AGe	TTG	CTb	GTb	TAb	CCd
Others	GGb			GAe	AAe	GCd	ACb	CAe	AAe	TCb	ATb	CAA	AAe	TTb	ACe		
	AGb						AAe										
							AGb										

Note 1: The first line shows numbers (italicized) of the amino acid residues for all the molecules, including gaps; the other numbers indicate the number of residues in the four respective chains. The lower half of the table shows the corresponding coding triplets for the amino acids in terms of the strand of DNA with polarity identical to that of messenger RNA. The third base in each coding triplet is abbreviated as follows: b = C or T; d = A, G, C, or T; e = A or G. * = invariant sites.

Table 5-2 (*continued*)

Residue No.:		*40*										*50*							
Protein		★	40			★		45								53			
Myo	glu	thr	leu	glu	lys	phe	asp	lys	phe	lys	his	leu	lys	ser	glu	asp	glu	met	lys
							44		46			48							
α-Hb	thr	thr	lys	thr	tyr	phe	pro	his	phe	…	asp	leu	ser	his					
									45				49			52			
β-Hb	try	thr	glN	arg	phe	phe	glu	ser	phe	gly	asp	leu	ser	thr	pro	asp	ala	val	met
γ-Hb	try	thr	glN	arg	phe	phe	asp	ser	phe	gly	asN	leu	ser	ser	ala	ser	ala	ilu	met
Others	ala		arg				ala	arg	leu	ala	gly	met	glu	gly		glu		leu	
									try	glu		ilu	thr			ala			
Myo	GAe	ACd	CTe	GAe	AAe	TTb	GAb	AAe	TTb	AAe	CAb	TTG	AAe	AGb	GAe	GAb	GAe	ATG	AAG
α-Hb	ACe	ACd	AAe	ACe	ATb	TTb	CCd	CAb	TTb	…	GAb	TTG	AGb	CAb					
β-Hb	TGe	ACd	CAe	AGe	TTb	TTb	GAe	AGb	TTb	GGe	GAb	TTG	AGb	ACb	CCe	GAb	GCe	GTA	ATG
γ-Hb	TGe	ACd	CAe	AGe	TTb	TTb	GAb	AGb	TTb	GGe	AAb	TTG	AGb	AGb	GCe	AGb	GCe	ATA	ATG
Others	GCe		AGe				GCd	AGe	CTb	GCe	GGb	ATG	GAe	GGb		GAe		TTG	
									TGe	GAe		ATA	ACb			GCd			

			60				*		*60*		*70*						*68*	*74*
Residue No.: Protein			54				63		60				69					
Myo	ala	ser	glu	leu	lys	lys	his	gly	val	thr	leu	ala	thr	ala	leu	gly	ala	ilu
α-Hb	gly	ser	ala	asp	val	lys	his	gly	lys	lys	val	ala	asp	ala	leu	thr	asN	ala
				gIN														
β-Hb	gly	asN	pro	lys	val	lys	ala	his	gly	lys	lys	val	leu	gly	ala	phe	ser	asp	gly
γ-Hb	gly	asN	pro	lys	val	lys	ala	his	gly	lys	lys	val	leu	thr	ser	leu	gly	asp	ala
Others	lys		phe	arg	arg	ala	asp	tyr	ala	arg	glu	glu	glu	gly	val	asp	lys	leu	glu
				gly		glu	try	arg	thr		ala	ilu	his	asp		asp	leu	glu	asp
				glu					arg				ser					glu	

Protein																	
Myo	GCe	AGb	GAe	GAb	CTb	AAe	CAb	GGd	GTe	ACe	GTA	CTb	ACb	GCd	CTb	GGb	GCb	ATb
α-Hb	GGe	AGb	GCd	CAe	GTb	AAe	GGd	CAb	AAe	AAe	GTA	GCb	GAb	GCd	CTb	ACb	AAb	GCd
β-Hb	GGe	AAb	CCd	AAe	GTb	AAe	GCd	CAb	GGd	AAe	AAe	GTA	CTb	GGb	GCd	TTb	AGb	GAb	GGd
γ-Hb	GGe	AAb	CCd	AAe	GTb	AAe	GCd	CAb	GGd	AAe	AGe	GAA	CTb	ACb	TCd	CTb	GGb	GAb	GCd
Others	AAe		TTb	AGe	TTb	AGe	TAb	GCd	GAe	GCe	AAb	ATA	GAe	GGd	GTb	GAb	AAe	GAe
				GGe		GCe	TGe	CGb		GAe			CAb	GAe			CTb	GAb
				GAe						ACe			AGb	GAb				
										AGe			GAe					

Note 2: The helical regions described by Perutz (1965) correspond to the following residues for all molecules: Region A, residues *4 to 19*; B, *21 to 37*; C, *38 to 43*; D, *52 to 58*; E, *59 to 78*; F, *87 to 95*; G, *101 to 119*; H, *126 to 150*. Perutz lists the other regions as follows: NA, *1 to 3*; AB, *20*; CD, *44 to 51*; EF, *79 to 86*; FG, *96 to 100*; GH, *120 to 125*; HC, *151 to 155*.

Table 5-2 (*continued*)

Residue No.: Protein		80					84					90*						
Myo	leu	lys 71	lys	lys 73	gly	his	his	glu	ilu 78	glu	leu	lys	pro 82	leu	ala	glN 85	ser	his 87	ala
α-Hb	val	ala	his	val	asp	asp 80	met	pro	asN	ala	leu	ser	ala	leu	leu	asp 90	leu	his	ala
β-Hb	leu	ala	his	leu	asp	asN	leu	lys	gly	thr	phe	ala	thr	leu	ser	glu	leu	his	cys
γ-Hb	ilu	lys	his	leu	asp	asp	leu	lys	gly	thr	phe	ala	glN	leu	ser	gly	leu	his	cys
Others	thr	gly				asN			ala		ilu		asN			thr		tyr	gly
		glu									tyr		lys						
		pro											gly						
		his											ser						
Myo	CTb	AAe	AAe	AAe	GGb	CAb	CAb	GAe	ATb	GAe	CTb	AAe	CCb	CTd	GCd	CAe	CTe	CAb	GCb
α-Hb	GTb	GCd	CAb	GTe	GAb	GAb	ATG	CCe	AAb	GCe	CTb	TCe	GCd	CTd	TCd	GAb	TTe	CAb	GCb
β-Hb	CTb	GCd	CAb	TTe	GAb	AAb	CTG	AAe	GGb	ACe	TTb	GCe	ACd	CTd	TCd	GAe	TTe	CAb	TGb
γ-Hb	ATb	AAe	CAb	TTe	GAb	GAb	CTG	AAe	GGb	ACe	TTb	GCe	CAe	CTd	TCd	GGd	TTe	CAb	TGb
Others	ACb	GGd				AAb			GCb		ATb		AAb			ACb		TAb	GGb
		GAe									TAb		AAe						
		CCd											GGd						
		CAb											AGb						

Residue No.: Protein					100							107	108	110				113
Myo	thr	lys	his	lys	ilu	pro	ilu	lys	tyr	leu	glu	phe	glN	ser	glu	ala	ilu	ilu	ser
α-Hb	his	lys	leu	arg	val	asp	pro 100	val	asN	phe	lys	leu	leu	ser	his	cys	leu	leu	val
β-Hb	asp	lys	leu	his	val	asp	pro	glu	asN	phe	arg	leu	leu	gly	asN	val	leu	val	cys
γ-Hb	asp	lys	leu	his	val	asp	pro	glu	asN	phe	lys	leu	leu	gly	asN	val	leu	val	thr
Others	val	glu						glN	asp		leu	ilu	val ilu		asp	ser		ala ser	leu asp

Myo	ACb	AAe	CAb	AAe	ATb	CCb	ATb	AAe	TAb	TTe	GAe	TTb	CAe	AGb	GAe	GCb	ATb	ATb	TCb
α-Hb	CAb	AAe	CTb	CGd	GAb	CCb	GTe	AAb	TTb	AAe	CTb	AGb	CAb	TGb	CTb	CTb	CTb	GTb
β-Hb	GAb	AAe	CTb	CAb	GTb	GAb	CCb	GAe	AAb	TTb	AGe	CTb	CTb	GGb	AAb	GTb	CTb	GTb	TGb
γ-Hb	GAb	AAe	CTb	CAb	GTb	GAb	CCb	GAe	AAb	TTb	AAe	CTb	CTb	GGb	AAb	GTb	GTb	GTb	ACb
Others	GTb	GAe						CAe	GAb		TTe	ATb	GTb ATb		GAb	CTb		GCb TCb	CTb GAb
	TCe																		

Table 5-2 (continued)

Residue No.: Protein					120								125	126	
Myo	val	leu	asN	ser	lys	his	pro	gly	asN	phe	gly	ala	asp	ala	glN
α-Hb	thr	leu	ala 110	ala	his	leu	pro	ala	glu 121	phe	thr	pro	ala 125	val 126	his
β-Hb	val	leu	ala	his	his	phe	gly	lys	glu	phe	thr	pro	pro	val	glN
γ-Hb	val	leu	ala	ilu	his	phe	gly	lys	glu	phe	thr	pro	glu	val	glN
Others	glu	val ser	his	arg val	asN arg			asN his	glN asp	lys	ser asp		glN	met leu phe	

Myo	GTe	CTb	AAb	TCb	AAe	CAb	CCd	GGd	AAb	TTb	GGb	GCd	GAb	GCG	CAe
α-Hb	ACe	CTb	GCb	GCb	CAb	CTb	CCd	CGd	GAe	TTb	ACb	CCd	GCd	GTG	CAb
β-Hb	GTe	CTb	GCb	CAb	CAb	TTb	GGd	AAe	GAe	TTb	ACb	CCd	CCd	GTG	CAe
γ-Hb	GTe	CTb	GCb	ATb	CAb	TTb	GGd	AAe	GAe	TTb	ACb	CCd	GAe	ATG	CAe
Others	GAe	GTe TCe	CAb	CGb GTb	AAb AGe			AAb CAb	AAe CAe GAb	AAe	TCb AAb		CAe	CTG TTb	

Residue No.:	130				*			130		140				142	
Protein															
Myo	gly	ala	met	asN	lys	ala	leu	glu	leu	phe	arg	lys	asp	met	ala
α-Hb	ala	ser	leu 130	asp	lys	phe	leu	ala	ser	val	ser	thr	val 140	leu	thr
β-Hb	ala	ala	tyr	glN	lys	val	val	ala	gly	val	ala	asN	ala	leu	ala
γ-Hb	ala	ser	try	glN	lys	met	val	thr	gly	val	ala	ser	ala	leu	ser
Others			phe		glN		phe	ser	thr, ala		thr	asp		ilu	
Myo	GGd	GCb	ATG	AAb	AAe	GCd	CTb	GAe	TTe	TTb	CGb	AAe	GAb	ATG	GCb
α-Hb	GCd	TCb	TTG	GAb	AAe	TTb	CTb	GCd	TCe	GTb	TCb	ACd	GTb	TTG	ACb
β-Hb	GCd	GCb	TAb	CAe	AAe	GTd	GTb	GCd	GGe	GTb	GCb	AAb	GCb	TTG	GCb
γ-Hb	GCd	TCb	TGG	CAe	AAe	ATG	GTb	ACd	GGe	GTb	ACb	AGb	GCb	TTG	TCb
Others			TTb		CAe		TTb	TCd	ACe, GCe			GAb		ATA	

Table 5-2 (continued)

Residue No.: Protein	144	145	*		150		151		154
Myo	ser	asp	tyr	lys	glu	leu	gly	tyr	glN	gly—COOH
α-Hb	ser		tyr	lys		arg—COOH			
				141					
β-Hb	his		tyr	lys		his—COOH			
						146			
γ-Hb	ser		tyr	arg		his—COOH			
						146			
Others	ala							phe	
	asp								
Myo	TCb	GAb	ATb	AAe	GAe	CTd	GGd	ATb	CAe	GGd TAe
α-Hb	TCb	AAe	ATb	CGd	TAe					
β-Hb	CAb	AAe	ATb	CAb	TAe					
γ-Hb	TCb	AGe	ATb	CAb	TAe					
Others	GCb							TTb		
	GAb									

Sources: Allan et al. (1965); Braunitzer (1965); Braunitzer et al. (1964); Clegg et al. (1965); Edmundson (1965); Hill et al. (1965); Jones et al. (1965, 1966); Perutz et al. (1965); Schneider and Jones (1965); Watson-Williams et al. (1965).

stimuli. Indeed, some of the mutations are deleterious, and in the ordinary course of evolutionary events, their possessors would have a diminished chance of survival.

6. Single-base changes to a code-terminating triplet account for the absence of the final six amino acids of myoglobin from α, β, and γ hemoglobins (GAA/TAA) and for the absence of the final six amino acids of myoglobin from lamprey hemoglobin (AAA/TAA) (Perutz et al., 1965).

The hemoglobin family of proteins has been studied to a greater extent and in more detail than any other group of homologous proteins. Hemoglobin is readily obtained in pure form and in large quantities from red blood cells, which contain most of their soluble protein as hemoglobin and are not nucleated. The importance of the anemias gives added impetus to the study of hemoglobin. The formation of hemoglobin in reticulocyte ribosomes made it, in the hands of Borsook, Schweet, and their collaborators, the first protein to be produced by an extracellular reaction taking place in vitro (Kruh and Borsook, 1956; Borsook, Fischer, and Keighley, 1957; and Schweet, Lamfrom, and Allen, 1958).

The study of hemoglobin has ramified in other directions. Pauling et al. (1949) discovered a difference in electrophoretic mobility between normal hemoglobin and hemoglobin from patients with the hereditary disease sickle-cell anemia. The difference was eventually shown to be due to a change of one amino acid in the β chain (Ingram, 1957), ascribable in turn to a change of a single AT pair to TA in the human genome (Ingram, 1958). This observation had a profound effect on molecular concepts in genetics.

Hemoglobin S

The findings with hemoglobin S, the abnormal hemoglobin that occurs in sickle-cell anemia, have been reviewed several times (Baglioni, 1963; Ingram, 1963), and their scientific implications are so extensive that they will again be summarized.

Sickle-cell anemia was described by Herrick in 1910 (see Ingram, 1963). The name is derived from the shape of the red blood cells occurring in carriers of the trait of the disease. The hemoglobin in the cells precipitates when it is deoxygenated and this produces crystalloid aggregates that distort the cells. The disease is rare in most countries but occurs commonly in central Africa and Madagascar and is found less commonly but still quite frequently in the countries surrounding the Mediterranean. Since the disease is harmful or lethal, it is surprising that it has not tended to disappear because of the elimination of its carriers. It has been suggested that its perpetuation is favored by the fact that individuals heterozygous for hemoglobin S are more resistant than normal individuals to malaria (Allison, 1954, 1964) and that selection in favor of the sickle-cell allele started when malaria increased as a result of the new breeding places for mosquitoes that appeared when the tropical forests were cleared. The increase in human population has also favored the malaria parasite by providing host material. These conditions would favor the selective survival of individuals carrying the sickle-cell trait.

Two modifications of the disease exist: in the first, carriers of the characteristic sickle-shaped erythrocytes are apparently healthy; in the second, the individuals have a severe hemolytic anemia. The red cells of both types contain an abnormal variant of the β chain of hemoglobin, termed β^S. Pauling (1953–54) designated sickle-cell anemia as a "molecular disease," a concept that was reinforced by the further characterization of the abnormality.

The two types of the disease were shown to be the heterozygous and homozygous forms, respectively. The red cells of heterozygous individuals contain normal hemoglobin A, which consists of two normal α chains and two normal β chains, so that it is written as $\alpha_2^A \beta_2^A$, and hemoglobin S, in which both β chains are abnormal ($\alpha_2^A \beta_2^S$). The mixed tetramer $\alpha_2 \beta^A \beta^S$ does not occur (Itano, 1960). The red cells of heterozygous patients do not contain equal quantities of $\alpha_2^A \beta_2^A$ and $\alpha_2^A \beta_2^S$; the ratio varies from 1:0.5 to 1:0.8. This ratio is most interesting from the standpoint of genetic control mechanisms, and several explanations are possible; for example, the mutant messenger RNA for β^S may be produced in smaller quantities than β^A messenger, or β^S polypeptide may be released more slowly than β^A from the polysomes. These points are considered below. Homozygous individuals will have red cells in which hemoglobin A is replaced entirely by hemoglobin S.

The procedure used by Ingram to identify the difference between hemoglobins A and S was as follows: Hemoglobin A and hemoglobin S were both digested with trypsin, which splits polypeptides at the junctions between lysine or arginine residues and the amino acid held in peptide linkage by the carboxyl group of lysine or arginine. Each of the four hemoglobin chains in hemoglobins A or S is broken into 14 or 15 polypeptide fragments of varying size by this procedure. Ingram found that one of the peptides in hemoglobin S could be distinguished from its counterpart in hemoglobin A by a difference in electrophoretic mobility. These two peptides were then analyzed for their content of amino acids. The peptide from hemoglobin S was found to contain one more valine and one less glutamic acid than the corresponding peptide from hemoglobin A. Later it was shown that the substitution had occurred in position 6 of the polypeptide chain of β hemoglobin, numbering from the end that is occupied by a free NH_2 group (Ingram, 1959). This seemingly trivial change, located at a long distance from the active center which combines with hemin in the globin molecule, was sufficient to alter drastically the properties of hemoglobin.

The physicochemical changes that were brought about by the substitution were subsequently explored by Murayama (1957). They are characterized by the following effects: Deoxygenated sickle-cell hemoglobin formed a gel at 38 C, but the gel liquefied on cooling to 0 C. Gel formation was prevented by dialysis or by oxygenation. The use of optical rotatory dispersion to measure conformational changes led Murayama (1962) to conclude that the valyl residue in position 6 of the β chain of hemoglobin S interlocked at 38 C with the amino-terminal valyl residue in position 1, thus allowing cyclization from carboxyl of the first valyl to the NH of threonyl at residue 4 by hydrogen bonding.

In a later report by Murayama (1965a), sickled erythrocytes were found to orient themselves perpendicularly to the lines of force in a powerful magnetic field. Deoxygenated erythrocytes containing hemoglobin A showed no such orientation. The effect was presumed to be due to the stacking of hemoglobin S molecules along the long axes of sickled erythrocytes. Murayama (1965b) has recently prepared electron photomicrographs of the long molecular "cables" of hemoglobin S.

The observations with hemoglobin S are a striking demonstration of the remarkable effects that can be produced by a single-base change affecting the code for a key amino acid in a protein molecule. Such effects emphasize the possibility for sudden changes occasionally appearing as the result of a mutation.

The formation of the tetrameric molecules of the various hemoglobins by combination of four completed polypeptide chains is a matter of much interest from the standpoint of genetics and biochemistry. Normal adult human blood contains the two hemoglobins A ($\alpha_2^A\beta_2^A$) and A_2 ($\alpha_2^A\delta_2^A$) in a proportion of about 25:1. Hemoglobin F ($\alpha_2^A\gamma_2^F$) is found in fetal blood instead of hemoglobin A, which replaces hemoglobin F during the first six months of neonatal life. Mutations involving the α and β chains at a single site, such as hemoglobins I and S, occur as $\alpha_2^I\beta_2^A$ and $\alpha_2^A\beta_2^S$, i.e., in the general form $\alpha_2^X\beta_2^Y$, not $\alpha^X\alpha^W\beta_2^Y$ or $\alpha_2^X\beta^Y\beta^Z$. A genetic cross between individuals carrying $\alpha_2^I\beta_2^A$ and $\alpha_2^A\beta_2^S$ hemoglobins would give rise to offspring carrying $\alpha_2^I\beta_2^A$, $\alpha_2^A\beta_2^S$, $\alpha_2^A\beta_2^A$, and $\alpha_2^I\beta_2^S$ hemoglobins but not combinations such as $\alpha^A\alpha^I\beta_2^A$, etc. In certain blood dyscrasias, β_4 (hemoglobin H) or γ_4 (hemoglobin Barts) molecules occur, but α_4 molecules have not been found in vivo, although they may be produced in vitro by suitable dissociation and combination of the appropriate units.

The regulation of hemoglobin synthesis in immature red cells which leads to the various proportions of the completed molecules must take into account the following observations:

1. Hemoglobins A and A_2 each contain two identical α chains but A is produced much more abundantly than A_2, although A_2 is a normal hemoglobin and its δ peptide chains differ in only 8 or 10 amino acid loci from β chains. Genetic studies indicate that the β and δ genes are closely linked (Huisman, Punt, and Schaad, 1961).

2. The replacement of the γ chain by the β chain occurs in the first

few months of life, so that fetal hemoglobin F ($\alpha_2^A \gamma_2^F$) is replaced by adult hemoglobin A ($\alpha_2^A \beta_2^A$). The genes for the β and γ chains are separate, and it appears as if the β gene is turned on while the γ gene is being turned off. This change was found to be accelerated in frogs by administering thyroxine (Moss and Ingram, 1965).

An illuminating article by Winterhalter and Huehns (1964) described the properties of $\alpha\beta$ globin subunits. Heme was removed from hemoglobin by acid acetone, and the resultant globin was found to consist of two polypeptide chains with the structure of $\alpha\beta$. When mixtures of $\alpha\beta$ globins were reconverted to hemoglobin, only identical $\alpha\beta$ subunits combined with each other; for example, a mixture of globins $\alpha^A\beta^C$ and $\alpha^A\beta^A$ gave hemoglobins $\alpha_2^A\beta_2^C$ and $\alpha_2^A\beta_2^A$. These findings lead to a suggestion that during protein synthesis an α globin chain is not released from the ribosome until it has combined with a β, γ, or δ chain. The β, γ, and δ chains are freely released from the ribosomes prior to dimer formation; in this regard it is noteworthy that they terminate with histidine-COOH, while the α chain terminates with arginine-COOH. The dimer ($\alpha\beta$, $\alpha\gamma$, or $\alpha\delta$) then combines with another identical dimer and four heme groups to form a complete hemoglobin molecule. The 25:1 ratio of HbA to HbA$_2$ and the replacement of HbF by HbA would therefore depend only on the relative rates of synthesis of the β, γ, and δ chains, which could be related in turn to the relative numbers of polysomes carrying messenger RNA for the β, γ, and δ chains. The amounts of these would regulate α-chain production at the polysome level, since the release of α chains would be in response to dimer formation with one of the other chains, and this would account for the absence of α_4 chains in vivo. The presence of β_4 and δ_4 chains in certain blood dyscrasias would be due to a lack of polysomes supplying α chains.

These considerations do not provide an explanation for the differences and variations in rates of production of the β, γ, and δ chains. Ingram (1963, pp. 90, 117), Baglioni (1963), and Zuckerkandl (1964) have proposed regulatory mechanisms for these based on the operon model of Jacob and Monod (1961), which relates to the *E. coli* galactosidase system. The cells of mammals and other higher organisms differ markedly from bacterial cells with respect to their organization of genetic material. Extending the operon theory to include mammalian cells involves an excursion which depends on various theoretical assumptions. The means by which gene expression is controlled in the cells of higher organisms are largely unknown and are the subject of much research.

The tertiary structure of myoglobin is very similar to that of the individual units of the hemoglobins. It is shown in highly simplified form in Fig. 5-3, which represents the orientation of the polypeptide chain. This tertiary structure is of strong evolutionary significance, since it depends on a secondary structure consisting of eight α helical segments, each containing from 6 to 25 amino acids joined together by nonhelical regions. The secondary structure depends in turn on a primary structure consisting of a sequence of 141 to 154 amino acids in peptide linkage. There are many differences between the amino acids at corresponding loci in the hemoglobins and myoglobins from various species, but a common pattern exists in certain respects: when the peptide chains of human myoglobin and the four human hemoglobins are compared, each of 21 widely scattered homologous sites is occupied in five protein chains by identical amino acids. This indicates that the hemoglobins and myoglobins have evolved from a single archetypal precursor molecule. However, more than half of these sites are changed in other species.

It was proposed by Zuckerkandl and Pauling (1962) that information about the archetypal molecule may be gained through comparison of homologous polypeptide chains, and an approach was made to this possibility by comparing the primary structures of the hemoglobins (Jukes, 1963a). An analogous attempt was made (Jukes, 1965b) by attempting to back-translate the various chains in terms of the coding triplets for the respective amino acids at each homologous residue; but the information on the code was at that time incomplete, and in any case the conclusions tend to be ambiguous. Such speculations will be more effective when it becomes possible to calculate the secondary and tertiary structures of proteins from an examination of the amino acid sequence.

Evolution and the Hemoglobins   171

We shall now consider the information on the composition of the various globins in terms of the amino acid code. This has been attempted in Table 5-2, which lists in the first four lines the amino acid sequences in human myoglobin and α, β, and γ hemoglobins, starting at the amino-terminal end of the chains. The italicized numbers above the amino acid residues in the top row (and as used in the text of this chapter) represent the combined overall number of loci in all the chains, counting all gaps. Other numbers are interspersed below to indicate the distance of residues from the starting point of the respective chains. Below the sequence of γ hemoglobin are listed various amino acids that correspond to mutations, δ hemoglobin, and the hemoglobins of other species, as further explained in Table 5-3.

The DNA codes corresponding to the respective amino acids are shown in the lower portion of Table 5-2. For convenience in making comparisons with messenger RNA triplets, these have been written in the form of the DNA chain with the same polarity as the messenger. In most cases the terminal base of the triplet can consist of either of two bases or of any of four bases. These are abbreviated as follows: b = C,T; d = A,G,C,T; e = A,G. In some cases a selection has been made between two different codes for the same amino acid with a view to minimizing the differences between codes at the same locus. For example, at residue *99* arg code CGb corresponds to his, but at residue *106*, arg code AGe corresponds to lys.

The sequences of bases in the lower lines of the table are an attempt to represent the genes corresponding to the four proteins under discussion and are written with the T of DNA instead of the U of RNA. In nearly every case it is not possible to specify the third base of a triplet because of the equivalence of more than one base in such positions. Exceptions have been made to this in loci that include methionine, which apparently is coded only by AUG. It has also been assumed that a locus containing ilu and glu should be represented by ATA for ilu and GAA for glu; this minimizes the number of base changes postulated.

Another source of ambiguity is caused by the multiplicity of codes for arginine, serine, and leucine. At numerous sites it is not possible

Evolution and the Hemoglobins

Table 5-3. Amino Acid Interchanges of Human Myoglobin and the α, β, and γ Chains of Hemoglobin When Compared with the δ Chain, with Variants, and with the Corresponding Chains of Other Species

Site Number	Position in Chain and Amino Acid		Amino Acids in Variants or in Other Species	Site Number	Position in Chain and Amino Acid		Amino Acids in Variants or in Other Species
1	M-1	gly	w-val	30	α-29	leu	r-ilu
	β-1	val	c-acetyl val; le-thr	31	α-30	glu	mu-glN
				32	αM-31	arg	la-lys
2	β-2	his	c-thr; b, h-glN; le-leu; mu-tyr	33			x-phe
				35			x-thr
	δ-2	his	δ mu-arg	36	M-35	gly	w-ser
3			x-ser		α-35	ser	h, r-gly
4	β-4	thr	h, r, le-ser	39	M-38	glu	la-ala
5	α-4	pro	h, la-ala	41			x-arg
	β-5	pro	h-gly; p-ala; r-ser	43	M-42	lys	la-phe
				45	M-44	asp	la-pro
	α-5	ala	mu-asp		α-44	pro	ca-ala
6	β-6	glu	mu-val, lys		β-43	glu	mu-ala
			x-asp	46	M-45	lys	w-arg
7	α-6	asp	la-glu	47	α-46	phe	ca-try
	β-7	glu	mu-lys, gly		β-45	phe	mo-leu
8	β-8	lys	le-asp	48	α-gap 46/47		ca-ala
9	β-9	ser	δ-thr; h-ala		β-46	gly	mu-glu
10	β-10	ala	le-his	49	M-48	his	la-gly
			x-gly, asp, lys		α-47	asp	mu-gly
11	α-10	val	r-ilu	50	M-49	leu	la-met
	β-11	val	h-leu		α-48	leu	r-ilu
12	β-12	thr	δ-asN; h-leu	51	M-50	lys	la-thr
			x-gly, asp, arg		β-49	ser	mo-glu
14	α-13	ala	r-glu		α-49	ser	r-thr
	β-14	leu	δ-arg	52	M-51	ser	w-thr
15			x-phe		α-50	his	ca-pro (or gly)
16	α-15	gly	h-ser; r-thr; mu-asp		β-50	thr	δ-ser; h-gly
				53	M-52	glu	la-ala
	β-16	gly	mu, h-asp;		α-51	gly	ca-gly (or pro)
	δ-16	gly	δ mu-arg	54	M-53	asp	w-ala
			x-lys		β-52	asp	mo-glu
17	α-16	lys	mu-glu; le-asp	56	M-55	met	la-leu
18			x-leu, ilu	58	M-57	ala	la-lys
20	M-19	pro	w-ala	60	M-59	glu	la-ala
	α-19	ala	r, h-gly		α-53	ala	ca-pro; r-phe, le-gly
			x-ser				
22	β-20	val	h-glu	61	α-54	glN	mu-arg, glu; ca-gly
23	α-22	gly	mo, mu-asp				
	β-21	asp	h-glu	62	M-61	leu	la-val
			x-lys				x-phe
24	α-23	glu	g-asp	63	M-62	lys	la-arg
	β-22	glu	δ-ala; δ mu-glu; mu-lys		β-61	lys	ca-ala, mu-glu
				64	M-63	lys	la-try
25	α-24	tyr	le-thr		α-57	gly	ou-asp; h, r-ala; mu-asp
27	M-26	asp	w-glN				
	α-26	ala	r-ser	65	β-63	his	mu-tyr, arg
28	β-26	glu	mu-lys		α-58	his	mu-tyr
29	α-28	ala	le-ser				

Table 5-3 (*continued*)

Site Number	Position in Chain and Amino Acid	Amino Acids in Variants or in Other Species	Site Number	Position in Chain and Amino Acid	Amino Acids in Variants or in Other Species
66	M-65 gly	la-ala	97	β-95 lys	mu, p-glu
	α-59 gly	r-ala		M-96 lys	la-ser
	β-64 gly	r-ala	103	M-102 lys	la-glN
67	M-66 val	la-glu	104		x-asp
	A-60 lys	b-glu; s-ala	106	β-104 arg	g-leu
	β-65 lys	ll-thr; ca-arg	107		x-ilu
68	M-67 thr	la-arg	108	M-107 glN	w-ilu
		x-asN			x-val
69	M-68 val	la-ilu	110		x-asp
	β-67 val	mu-glu	111		x-ser
70	M-69 leu	la-ilu	113	β-111 val	h-ala; le-ser
71	M-70 thr	la-asp	114	α-107 val	h-ser
	β-69 gly	mo-glu; h-his; le-ser		β-112 cys	h-leu; le-asp
				M-113 ser	w-his
	α-64 asp	r-glu	115		x-glu
72	α-65 ala	h-gly; r-ser	116	β-114 leu	h-val
	β-70 ala	h-ser			x-ser
		x-glu, asp	117	M-116 asN	w-his
73		x-val	118	α-111 ala	h-val; r-ser
74	β-72 ser	h-gly		β-116 his	δ, h-arg
		x-asp	119	β-117 his	δ-asN
75	α-68 asN	mu-lys; h-leu		M-118 lys	w-arg
	β-73 asp	h-glu	120	α-113 leu	r-his
76	β-74 gly	mo-glu, asp	122	α-115 ala	r, h-asN
77	β-75 leu	mo-val; thr; h-val			x-his
			123	α-116 glu	mu-lys; h-asp
78	α-71 ala	r, h-gly		β-121 glu	mu-lys, glN; h-asp
	β-76 ala	mo-glu, pro; h-his; p-lys	124		x-lys
80	α-73 val	h-leu	125		x-ser, asp
81	β-79 asp	mu-asN	127	β-125 pro	δ-glN; h-glu
83	α-76 met	h, r-leu	128	β-126 val	δ-met; h-leu
85	M-84 ilu	w-ala			x-phe
	α-78 asN	h-gly	131	β-129 ala	h-ser
87		x-ilu, tyr	132		x-phe
88	β-87 ala	δ-ser	134	β-132 lys	mu-glN
89	α-82 ala	h-asN; r-gly	136		x-phe
	β-87 thr	mu, p-lys; δ-glN, h-ala	137	α-130 ala	h-ser
			139		x-thr, ala
		x-ser	140	α-133 ser	r-thr
92	α-85 asp	r-thr	141	β-139 asN	le-asp
	β-90 glu	ll-glN	143	M-142 met	w-ilu
94	α-87 his	mu-tyr	145	M-144 ser	w-ala
95	α-88 ala	r-gly		β-143 his	mu-asp
96	β-94 asp	le-val	146	M-145 asp	w-lys
			152	M-151 tyr	w-phe

w = whale; c = chicken; b = bovine; h = horse; mu = human mutants; δ = human; g = gorilla; r = rabbit; ca = carp; mo = monkeys (various); ou = orang-utan; p = pig; s = sheep; la = lamprey; ll = llama; le = *lemur fulvus*; x = undesignated, from Perutz et al. (1965).

Note: Site numbers correspond to the italicized numbering in Table 5-2.

Sources: Allan et al. (1965); Braunitzer (1965); Braunitzer et al. (1964); Clegg et al. (1965); Edmundson (1965); Hill et al. (1965); Jones et al. (1965, 1966); Perutz et al. (1965); Schneider and Jones (1965); Watson-Williams et al. (1965).

to select between AGe and CGd for arg, between AGb and TCd for ser, or between CTd and TTe for leu. In other instances, however, a selection can be made on the strength of coding relationships to other amino acids at the same site, according to the scheme in Table 5-4. The amino acid sequences and proposed base sequences are schematically presented in Figs. 5-5 and 5-6. These make it possible to perceive

Table 5-4. Unique Single-Base Change Relationships for arg, leu, and ser Codes

Amino Acid	Code	Possible Mutations by Single-Base Changes
arg	AGe	ilu, lys, met, thr
	CGd	cys, glN, his, leu, pro
ser	AGb	arg, asN, gly, ilu
	UCd	ala, leu, phe, pro, try, tyr
leu	CUd	arg, glN, his, pro
	UUe	ser, try

b = U, C; d = A, G, U, C; e = A, G

the similarities and differences between the globin chains at a glance without wading through the complex mass of symbols in Table 5-2. Comparison of Figs. 5-5 and 5-6 shows that there is a closer relationship between the four genes than between the four proteins. The explanation is that the genes are in a four-letter language, but the proteins are in a 20-letter language, which tends to magnify the differences because of its greater versatility.

What are the third, or synonymous, bases in the triplets in the hemoglobin genes—are they predominantly C rather than T, or G rather than A? Table 5-2 indicates that the total of the bases in positions 1 and 2 of each coding triplet is almost equally divided between $G + C$ and $A + T$. If the hemoglobin genes have the same base composition as the average of human DNA, which is about 40 percent $G + C$ and 60 percent $A + T$, then about 80 percent of the triplets should end in G or C and 20 percent in A or T. An experimental examination of this suggestion is at present not possible. It is probable that the genes are well stabilized and that there is very little $A \leftrightarrows G$ and $T \leftrightarrows C$ fluctuation in the synonymous bases as they occur

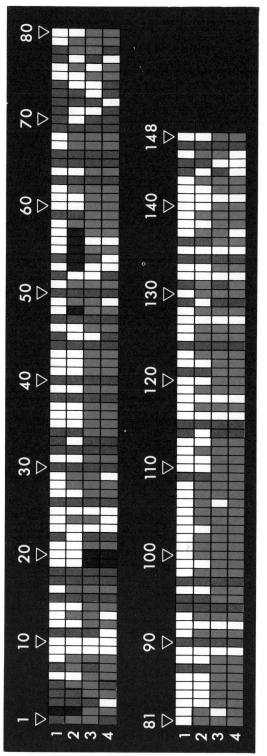

Fig. 5-5. Amino-acid replacements in human globin series: (1) myoglobin, (2) α hemoglobin, (3) β hemoglobin, (4) γ hemoglobin chains. Red areas indicate identity of hemoglobin sites with myoglobin; green areas identity of α, β, and γ hemoglobin sites with each other. Dark areas indicate gaps.

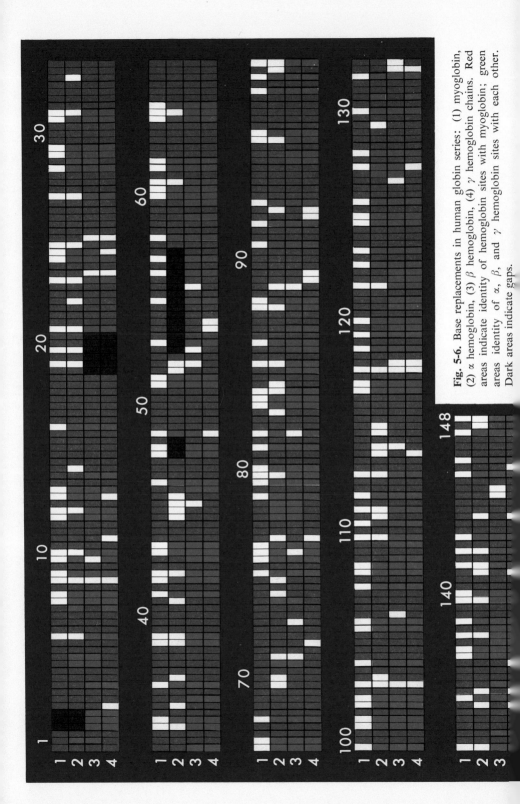

Fig. 5-6. Base replacements in human globin series: (1) myoglobin, (2) α hemoglobin, (3) β hemoglobin, (4) γ hemoglobin chains. Red areas indicate identity of hemoglobin sites with myoglobin; green areas identity of α, β, and γ hemoglobin sites with each other. Dark areas indicate gaps.

in the gene. The $A + T : G + C$ ratio of the DNA of mammals is relatively constant in the neighborhood of 58:42 (Chargaff, 1955), although the ultracentrifugal analysis of mammalian DNA shows an unsymmetrical distribution of regions of varying density, indicating that the $A + T : G + C$ ratio varies in different parts of the genome (Sueoka, Marmur, and Doty, 1959). This would be expected from the existence of specific regions for the transcription of rRNA, sRNA, and messengers for specific proteins. The most telling argument for constancy in the synonymous bases comes from the studies of cross-annealing hybridization between single strands of melted DNA obtained from closely related species reported by Bolton, McCarthy, and co-workers (Chapter 7). Extensive reshuffling of synonymous bases would be expected to interfere with hybridization of genetic material obtained from closely related species such as rats and mice, but such hybridization readily takes place.

The stabilization of the synonymous bases must also be related to the availability of various species of sRNA molecules. A shift in gly codes from GGU to GGA or GGG in messenger RNA presumably might need to be accompanied by a decrease in one species of glycyl sRNA and an increase in others, if protein synthesis is to be maintained at a normal rate.

The genes for the hemoglobin chains in Table 5-2 are viewed as being subject to mutational events occurring during evolution. The majority of these events are single-base changes, or "hits," distributed at random over the length of each strand. As the elapsed time increases, the distribution of hits will be such that an increasing number of coding triplets will receive more than one hit per triplet, as predicted by the Poisson distribution. The calculation of the probabilities involved as compared with the results observed is complex, because (1) any changes that produce nonfunctioning hemoglobin molecules will be lethal; and (2) changes in the third base of a coding triplet will often be imperceptible owing to synonymity. Nevertheless, the general principle may be stated that the number of two-base changes per one-base change in the total assembly of coding triplets in a globin gene will

176 Evolution and the Hemoglobins

increase with time (Jukes, 1963a). This is well borne out by the summary in Table 5-5. It must also be assumed that occasionally a base which has been changed once may be changed back to the original base as a result of a second mutational event. Such occurrences can be calculated statistically but cannot be perceived.

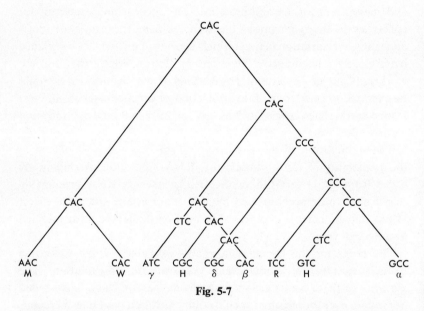

Fig. 5-7

What was the amino acid sequence of the archetypal "protoglobin"? This is impossible to write, and perhaps the effort to try to do so is not profitable in view of the large numbers of different sequences that currently exist in the hemoglobins of the animal world. Extinct evolutionary variations of these project backward into the dimmest mists of antiquity. Let us write a single site *118* in terms of evolutionary changes (Fig. 5-7). According to this speculation, the archetypal amino acid was histidine, which is found currently at this site in whale myoglobin and the β chain of human hemoglobin. Various changes shown in Fig. 5-7 have resulted in the same site being occupied by asN (human myoglobin), ilu (human γ), arg (horse β, human δ), ala (human α), and val (horse α). Pro and leu would be possible evolutionary intermediate stages in these pathways. Such theorizing, however, is subject to salutary restriction at the hands of protein chemistry; the hemoglobin molecule does not consist of a random series of amino acids but instead is a highly specified structure with limited flexibility of composition.

An examination and a comparison of the base sequences in Table 5-2 are summarized in Table 5-5. Twenty-three of the loci in the human globin chains are occupied by the same amino acid residue in all five positions. These have been regarded as "invariable" in establishing the homology of the five human globin chains with each other and in making the summary in Table 5-5, which compares the five chains (M, α, β, γ, δ). It is shown in Table 5-3, however, that 14 of these sites are variable in the sense that they are replaceable by other amino acids in different species. These are sites *15, 17, 18, 22, 30, 32, 38, 47, 50, 63, 66, 69, 97, 116,* and *124*. Are these replacements in other species due to the fact that such species have globin chains that differ in other respects from human chains? Has the replacement of val at *22* in the human globins by glu in horse β hemoglobin come about because of other differences between horse β and human hemoglobin?

The variation in these nine sites leaves only nine remaining invariant residues: *26* (gly), *38* (pro), *40* (thr), *44* (phe), *65* (his), *90* (leu), *94* (his), *134* (lys) and *147* (tyr). The specific functions of these were discussed by Perutz and coworkers as follows: the two histidines are linked to heme; leu, phe, and thr are in contact with heme; thr and pro are at the corner between the B and C helices; tyr is internally hydrogen-bonded with the α-CO group of the residue at site *100*; gly is at a cross-over point with gly or ala at residue *66* of the E helix; and lys at residue *134* has no known function. Are these nine truly invariant? Mutations of *65* and *94* are pathological changes and do not indicate true variability. Perhaps additional information will be furnished by globins which are evidently different in major respects from those that have been analyzed. Braunitzer and Braun (1965) have isolated four hemoglobins from the larvae of the dipterous insect *Chironomus thummi*. These contained

Table 5-5. Base Changes and Amino Acid Changes in the Relationship between Human α, β, γ, δ Hemoglobins and Myoglobin (M), between Human and Horse α and β Hemoglobins (α/h, β/h), and between Human and Whale Myoglobin (M/W)

	M/α	M/β	M/γ	α/β	α/γ	β/γ	β/δ	α/h	β/h	M/W
Variable loci	119	123	123	118	118	125	125	121	125	132
Apparent base changes in variable loci:										
None	14	15	15	43	38	85	115	105	99	116
One-base	57	53	57	53	55	30	9	12	18	11
Two-base	47	54	49	22	25	10	1	5	8	5
Three-base	1	1	2	0	0	0	0	0	0	0
Transitions	45	60	61	30	41	22	2	8	8	8
Transversions	109	104	100	67	65	28	9	14	26	13
Base changes[a]	0.43	0.44	0.44	0.28	0.30	0.13	0.029	0.060	0.090	0.053
Amino acid changes[a]	0.88	0.88	0.88	0.64	0.68	0.31	0.08	0.14	0.21	0.12

[a] Average per variable locus.

polypeptide chains of 124 to 127 residues and existed as $\alpha\beta$ dimers. The sequential analysis of these might reveal a further evolutionary stage in the hemoglobins.

It is remarkable that although the genes for the α and β chains show only 72 percent coincidence in terms of bases at the variable sites, each differs from the M gene by almost exactly the same amount, 43 or 44 percent. The same holds true when the α, γ, and M chains are compared. The same finding is reflected in the amino acid comparisons, although in this case the disparity between the hemoglobin chains and the myoglobin chain is heightened so that only 12 percent of coincidence remains. It is these comparisons that led to Ingram's conclusion that the hemoglobin and myoglobin genes have evolved from a single archetype in parallel pathways following gene duplications. Having accepted this conclusion, we are confronted with the remarkable fact that evolutionary divergence has changed the globin genes of a single species, *Homo sapiens*, from each other just as markedly as if they had been separated by speciation. Indeed, the difference between human myoglobin and α hemoglobin is quantitatively the same as the difference between whale myoglobin and human α hemoglobin. The divergence within the human species is so definite that the evolution of the duplicated genes within the cell can be plotted on the same time scale as that used for the divergence of the hemoglobins of the rabbit and the horse. It would appear that DNA is subject to random evolutionary events, most of which are base-pair changes, and that these events are distributed lengthwise along the DNA molecules. Presumably, only a tiny fraction of such changes reach the next generation of a higher organism through being carried into a fertilized ovum. Only a minute proportion of the changes carried to the next generation will have a perceptible effect on the total genetic pool of the species, so that after more than 100 million years of evolution only 22 base-pair differences are needed to account for the disparity between the α hemoglobin chains of horse and man.

Base substitutions that produce amino acid changes may be termed *code-altering* evolutionary changes. Myoglobin is separated from the hemoglobins by about 44 such changes per 100 base pairs, indicating

180 Evolution and the Hemoglobins

that about 22 code-altering changes per 100 base pairs have occurred during the evolution of the genes for myoglobin and α, β, and γ hemoglobins from the archetype, assuming that each gene has changed at approximately the same rate.

"Silent" changes, which do not produce an amino acid substitution, such as TTG,leu to CTG,leu, are not perceptible. Two-step changes that have restored an amino acid to its original locus, such as, perhaps, AAG/GAG/AAG, lys/glu/lys, leave no record in a comparison of two globin chains.

Myoglobin resembles the primitive lamprey hemoglobin in being single-stranded and in a number of its amino acid residues. It might be argued that α hemoglobin has changed more extensively from the ancestral protoglobin than has myoglobin. The available information does not support this argument: whale and human myoglobins have diverged from each other to the extent of about 5.3 percent base changes, and horse and human α hemoglobins to about 6.0 percent base changes, indicating that the myoglobins are evolving at about the same rate as the hemoglobins. Nevertheless, if the lamprey is indeed a "living fossil" and in its hemoglobin sequence, in common with its other characteristics, has therefore changed but slightly during evolutionary history, the closest clue to the amino acid sequence in the archetypal protoglobin might well be found by scrutinizing the amino acid sequences in the six lamprey hemoglobins (Rumen and Love, 1963). Complete information is not available for any of the lamprey hemoglobins, but a few of the differences are listed in Table 5-3.

Base changes apparently occur without respect to type, since the total number of transitions listed in Table 5-5 is 285 and of transversions is 535. The randomized chances for a transversional change are twice as great as those for a transition, since each purine can change to only one other purine but to either of two pyrimidines. For perfect randomness, the exact figures would be 273 transitions and 547 transversions in 820 changes. The proportion found in Table 5-5 is surprising in view of the emphasis that has been placed on the expected predominance of transitions by several authors, particularly in the case of chemical mutagenesis (see Chapter 4). The conclusion is that natural selection for the most useful type of protein must act to eliminate any chemical tendency that favors transitions over transversions if such a tendency exist.

Table 5-2 also indicates that only a small number of amino acid changes are produced by transversions in the third base of a coding triplet. Only 19 of about 149 loci show such changes, the most common being glu/asp and asN/lys. Transitions in a third base are not detectable except ilu/met.

The evolutionary separation of the hemoglobin genes is illustrated in Fig. 5-8, which is an elaboration of Ingram's diagram (Fig. 5-4).

The Hemoglobin Variants

The hemoglobin variants, or abnormal hemoglobins, are caused by a series of genetically determined changes in the human hemoglobin molecule. They are usually detected as a result of differences in electrophoretic migration rates from that of hemoglobin A. Many have been characterized chemically, and in every such case the substitution of a single amino acid, corresponding to a single-base change in a coding triplet, has been found to be responsible for the difference. It is quite possible that many variants have not been detected because they are the result of base substitutions that produce amino acid changes which do not affect the electrophoretic or other easily measurable properties of hemoglobin. On the basis of an examination of 120 individuals at random, Ingram (1963, p. 79) has pointed out that the probability of such hidden mutations is not large.

The nomenclature of the hemoglobin variants as commonly employed by hematologists presents a romantic geographical display, examples of which are $HbG_{Honolulu}$, HbD_{Punjab}, $HbD_{Turkish\ Cypriot}$, Hb Zurich, HbO_{Arabia}, and Hb Shimonoseki. At the expense of losing the international flavor of the series, they are listed in Table 5-6 solely under the designations of the amino acid substitutions involved. Note that all the substitutions are accompanied by a change in charge, thus accounting for the alterations in electrophoretic migration rates of the molecules.

The phenotypic properties conferred by the mutations vary greatly.

182 Evolution and the Hemoglobins

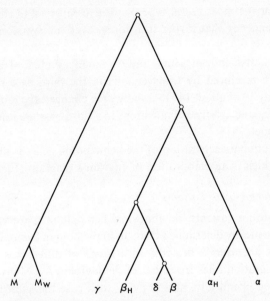

Fig. 5-8. Evolutionary separation of the genes for the globin chains, plotted on a time scale equivalent to the numbers of base changes (Tables 5-2 and 5-5). M = human myoglobin; Mw = whale myoglobin; α, β, γ, δ = human hemoglobins; α_H, β_H = horse hemoglobins; O = gene duplications.

Sickle-cell anemia hemoglobin S (β-6 val) has been discussed. The mutant hemoglobin C (β-6 lys) also causes a sickle-cell anemia with properties in the hemoglobin similar to but less deleterious than those of S. In addition to resistance to cerebral malaria, the S gene has been reported to cause increased fertility in mothers (Firschein, 1961) and to depress the rate of the synthesis of the β^S chain as compared with that of the normal β^A chain (Neel, Wells, and Itano, 1951). Indeed, Hill, Buettner-Janusch, and Buettner-Janusch (1963, 1965) regard the fertility effect of the S gene as being responsible for its widespread occurrence, and they discount the claim that there is any effect of resistance to malaria.

The hemoglobins D in five of seven cases differ from hemoglobin A

Table 5-6. Mutational Changes in Human Hemoglobin

Location and Amino Acid	Amino Acids in Variants	Base Change in Messenger RNA Triplets
Alpha chain:		
α-5, ala	asp	GCb/GAb
α-15, gly	asp	GGb/GAb
α-16, lys	glu	AAe/GAe
α-22, gly	asp	GGb/GAb
α-30, glu	glN	GAe/CAe
α-47, asp	gly	GAb/GGb
α-54, glN	arg, glu	CAe/CGe, GAe
α-57, gly	asp	GGb/GAb
α-58, his	tyr	CAb/UAb
α-68, asN	lys	AAb/AAe
α-87, his	tyr	CAb/UAb
α-116, glu	lys	GAe/AAe
Beta chain:		
β-2, his	tyr	CAb/UAb
β-6, glu	val, lys	GAe/GUe, AAe
β-7, glu	gly, lys	GAe/GGe, AAe
β-16, gly	asp	GGb/GAb
β-26, glu	lys	GAe/AAe
β-43, glu	ala	GAe/GCe
β-48, gl	glu	AAe/AAb
β-61, lys	asN	GGe/GAe
β-63, his	tyr, arg	CAb/UAb, CGb
β-67, val	glu	GUe/GAe
β-70 or 76, ala	glu	GCe/GAe
β-79, asp	asN	GAb/AAb
β-87, thr	lys	ACe/AAe
β-95, lys	glu	AAe/GAe
β-121, glu	glN, lys	GAe/CAe, AAe
β-132, lys	glN	AAe/CAe
β-143, his	asp	CAb/GAb
Gamma chain:		
γ-5 or 6, glu	lys	GAe/AAe
Delta chain:		
δ-2, his	arg	CAb/CGb
δ-16, gly	arg	GGb/CGb
δ-22, ala	glu	GCe/GAe

b = C, U; d = A, G, C, U; e = A, G

because of a change of β-121 from glu to glN. They occur frequently in India and induce no known abnormality except possibly a mild hemolytic anemia. Hemoglobin M is accompanied by methemoglobinemia. These are three different types of HbM: α-58 tyr, α-63 tyr, and α-67 glu. The change of his to tyr at α-58 or β-63, or val to glu at β-67, evidently causes binding between tyr (or glu) and Fe^{+++} of heme, so that a complex is produced which is not readily reduced, thus interfering with reversible oxygen transfer. In contrast, a nearby substitution, α-57 gly/asp, produces no abnormal behavior in hemoglobin.

Hemoglobin Zurich, β-63 arg, has the remarkable property of causing a hemolytic anemia in patients treated with sulfanilamide drugs (Neel et al., 1951). Ingram (1963, p. 75) points out that the variants with β-chain changes occur far more commonly than α-chain variants, perhaps because the latter may produce death *in utero*. Changes in the β chain are not expressed during prenatal life and early infancy, because at this time γ rather than β hemoglobin is in use.

The hemoglobin variants show us a remarkable picture of molecular evolution in action. It must be suspected that similar arrays of single-base changes in other body proteins are scattered broadcast throughout the human race, occurring sporadically in many parts of the world as in the case of the hemoglobins D, occasionally producing balanced polymorphism, as in the case of the accompanying effects caused by hemoglobins C and S, and perhaps fluctuating with the passage of time, as shown in the sequential evolution of wing patterns in another species, the scarlet tiger moth (Savage, 1963, p. 33).

The hemoglobin chains found in various species of animals may also be compared with each other with a view to searching for correlations between taxonomy, evolutionary separation, and amino acid substitutions. The known differences between such chains are shown in Table 5-3.

An examination of the amino acid sequences in the α and β hemoglobin chains of various species shows a number of changes, ranging in numbers from species to species in a manner roughly correlated with taxonomic classification. The information in the table is incomplete in the case of many of the species listed. Hill and the Buettner-Janusches (1963, 1965) have concentrated on an examination of the α and β hemoglobin chains in primates. They found that α chains have varied less than β chains during the evolution of primates, an observation paralleled by a comparison of the α and β hemoglobins of human beings and horses (Table 5-5). The β chains of lower primates were at many loci identical with human γ chains rather than with human β chains, again indicating that the β chain is of comparatively recent origin, perhaps having been separated by gene duplication only 60 million years ago. Some primates appeared to lack A_2-type hemoglobin, thus emphasizing the comparatively recent origin of the δ chain.

Table 5-3 shows some of the amino acid differences between myoglobin and α and β chains in humans and a few other species. This comparison is complete in the case only of whale myoglobin and horse α and β hemoglobins. In other cases the proteins are still under investigation, but it has been stated that the primary structure of lamprey hemoglobin (Braunitzer, 1965) contains no sequence gaps, thus further suggesting that it is primitive in its nature.

Can the large number and variety of amino acid changes, totaling about 700 when all the known globin chains in Tables 5-2 and 5-3 are compared, be placed in any kind of coherent pattern? The tertiary structure of the molecule is formed by eight helical regions of polypeptide chains connected by nonhelical regions of varying length. Evidently, the amino acids can vary greatly in both the helical and nonhelical regions provided that there are hydrophobic amino acid residues at the appropriate loci in both of these regions to stabilize the tertiary structure. The requirement for holding the heme group involves only a small part of the molecule. Hydrophobic interchanges occur at many sites. Many such interchanges are favored by the composition of the codes for hydrophobic amino acids. At other sites there are interchanges between hydrophobic and hydrophilic residues, between acid and basic residues, and between amino acids with large and small side chains. Perutz and coworkers (1965) have pointed out that the globins are much more flexible with respect to primary structure than are the cytochromes c, the difference being a reflection of the contrast in the tertiary structures involved. Globins are well supplied with α-helical segments which fold into the characteristic tertiary structure shown in Fig. 5-3. The interior of the chains consists predominantly of residues having hydrophobic side chains. These repeat on the average at intervals of about 3.6 residues along the α-helical segments, so that their interior faces are nonpolar. Replacements of many different kinds, including hydrophobic and hydrophilic interchanges, occur on the surface of the globin molecules and in surface crevices. Histidine, glutamic acid, or aspartic acid residues are a constituent of every corner, of its immediate neighborhood, and of every known helical region. Proline residues are present only in known helical regions or at the ends of helices; and in many instances serine, threonine, aspartic acid, or asparagine are at the first site in the amino end of the helix, followed by a proline at the second site. The great evolutionary adaptability of the molecule is evident.

The great variability of the globin molecule provokes another question: Was there only one archetypal globin molecule? If so, did it have a number of ancestors? The archetypal concept is based on the existence of diverging evolutionary pathways; these may be followed backward to their origin in a single progenitor molecule as indicated in Fig. 5-7. Speciation and divergent evolution have separated the descendants of this molecule. The preservation of information in DNA keeps the amino acid sequences of proteins ordered and intact in living organisms. It is appropriate, however, to speculate on the nature of enzymatically active molecules that possibly existed in "prebiological systems," meaning systems that existed prior to the appearance of cells. Did these follow a pathway of evolutionary convergence rather than divergence? This seems quite possible. As emphasized by Commoner (1965), catalytic activity may appear in a protein as a result of the generation of a rather wide range of amino acid sequences. Proteins or "proteinoids" of abiogenic origin, present in the "primitive soup," might have weak catalytic activity. The earliest organisms might tend to select and incorporate within their boundary membranes the most efficient of these proteins or "pre-enzymes." If some mechanism could develop for the organisms to perpetuate the "best" pre-enzymes, a process of "converging" evolution could start. Let us propose that at first the genetic code was in some way *read backwards from protein to nucleic acid*. The genetic coding process in its present form could not, of course, be translated in reverse. The very existence, however, of a machinery as complicated as the protein-synthesizing mechanism in cells is a striking demonstration of the intricacy that can be produced by biological systems. The essential component of a procedure for the translation of amino acid sequences into base sequences would be the recognition of polynucleotides by polypeptides. This is known to take place in the aminoacyl-sRNA synthetases and may also exist in RNA polymerase, and in the repressor and inducer molecules that recognize operator sites in DNA. The pre-enzymes that had been formed by random condensation of amino acids in "nonliving" chemical reactions could thus translate their amino acid sequences into base sequences, perhaps of RNA, which could be replicated by an RNA-synthetase type of procedure and retranslated into proteins. At a later stage, perhaps, the RNA itself could have been transcribed backwards into DNA; this is easy to visualize in view of the nature of the RNA-polymerase reaction, the known formation of DNA-RNA hybrid strands, and the influence of manganese on the selection of nucleotides. When the code is read backwards, the vexing question of the origin of the first nucleic acid informational macromolecules finds a solution which does not present the seeming insuperable enigma of the initial appearance of a genetic nucleic acid molecule carrying the coded information for biologically active proteins that were previously non-existent.

To return to hemoglobin, the concept that the amino acid substitutions in the globins originated in single-base changes seems unescapable. The evidence for this is as follows:

1. All the mutations in the hemoglobin variants represent single-base changes in coding triplets.

188 Evolution and the Hemoglobins

2. The most closely related chains, β and δ, differ by nine single-base changes and one two-base change.

3. As the disparity increases between the polypeptide chains, two-base changes in the coding triplets make their appearance in expected numbers on the basis of randomly distributed single-base changes throughout the genes.

4. At many loci intermediate steps in a multibase change are actually present, such as glu/ala/thr/ilu, GAA/GCA/ACA/ATA, site *29*; asN/asp/ala, AAb/GAb/GCb, site *75*; pro/ala/glu, CCd/GCd/GAe, site *127*. In other cases, it may reasonably be concluded that an intermediate amino acid has vanished during evolution. An example is β-76, human/horse, ala/his, GCb/CAb. This site is occupied in a monkey by pro, which is intermediate in its code, CCb.

Single-base changes predominate in the comparison, and, as might be expected, the most frequent are those involving the commonest amino acids in the globin molecules, such as ala/ser. The distribution of interchanges throughout the loci (Table 5-2) is quite well scattered (Table 5-7). It is difficult to examine this distribution statistically because the information on several of the animal chains is incomplete and because it includes the human variants, in which the only changes that have been detected are those in which a difference in charge has taken place. For 20 loci changed once in 100 sites, the Poisson distribution should be approximately as follows:

No. of changes:	0	1	2	3	4	5	6	7
Percent:	8	20	25	21	14	7	3	1

The changes thus tend towards a pattern of randomness with respect to both distribution and type. Of much interest is the average number of base changes per coding triplet per amino acid change in the four human globins as follows:

M/α	M/β	M/γ	α/β	α/γ	β/γ
1.46	1.50	1.50	1.29	1.32	1.25

These figures indicate that as the number of differences between the molecules increases, the individual differences between each homologous pair of coding triplets also show an average increase. The

Table 5-7

Number of Amino Acids at Locus	Number of Loci	Percent
1	6	4.0
2	29	20.0
3	49	33.0
4	32	22.0
5	19	13.0
6	8	5.4
7	3	2.0
8	2	1.4

agreement between the M/α, M/β, and M/γ comparisons and between the α/β and α/γ comparisons is remarkable.

Fitch (1966) has found that repetitive sequences are present in the α and β chains of human hemoglobin at an interval of 66 amino acid residues, indicating partial internal duplication in the ancestral globin gene. Our conclusions are that the region extending from *59* to *79* is homologous with the region from *125* to *145* (C. R. Cantor and T. H. Jukes, *in preparation*, 1966). The homology of these regions is most strongly marked in the β chain. There is also homology between *50* and *58* (assuming a gap between *52* and *58*), *70* and *79*, and *136* and *145*. Evidently, the primitive globin gene evolved by partial duplication of a still earlier gene.

All the conclusions support the concept that the evolutionary separation of two genes with a common origin may be measured in terms of base changes randomly distributed along strands of DNA. The great flexibility of composition in the globin polypeptide chains enables this concept to be thoroughly examined.

The red color of blood has excited the emotions and interest of man since the earliest records of history, yet never has the most imaginative of the literati conjured up a proposal that this incarnadine hue could reveal a story of the evolution of all animals, going back for hundreds of millions of years to a common ancestor.

The long and arduous tasks of the laboratory; the insight and skill of mathematicians, physicists, and chemists; the pooling and interweaving of scientific knowledge—all these unite to bring forth achievements which kindle an aesthetic satisfaction fully equal to that obtained by contemplating a major work of art. The exposition of the structure of the hemoglobin molecule is such an achievement. In its convolutions, helices, and bonds we see the changing interplay of natural forces and particles that give rise to the varied tapestry of life.

[6]

The Cytochromes c and Other Proteins Showing Evolutionary Changes

Let this work of selection on the one hand, and death on the other, go on for a thousand generations, who will pretend to affirm that it would produce no effect?
Charles Darwin, 1858
[J. Proc. Linnaean Soc. (London) 3:49 (1859)]

Cytochrome c

We have considered the case of the hemoglobin family of proteins mainly as an example of evolutionary changes in polypeptides that have taken place subsequent to the duplication of genes. The hemoglobins are also under study with respect to the amino acid sequences of homologous molecules occurring in different species, but the information in most cases is available only for parts of the molecules.

Knowledge of the primary structures of the cytochromes *c* in different species is far more extensive than in the case of the hemoglobins. The amino acid sequence of the "mammalian type" of cytochrome *c* was published for horse cytochrome *c* by Margoliash et al. (1961). It contained 104 amino acids. The cytochromes *c* of other species were found to be remarkably similar, but each included characteristic differences in its amino acids. The primary structure was found for the cytochromes *c* from man (Matsubara and Smith, 1963), beef (Yasunobu et al., 1963), chicken (Chan et al., 1963), tuna (Kreil, 1963), baker's yeast (Narita et al., 1963), dog (McDowall and Smith, 1965), rhesus monkey (Rothfus and Smith, 1965), rattlesnake (Bahl and Smith, 1965), *Neurospora* (Heller and Smith, 1965), *Candida krusei* (Narita and Titani, 1965), and a moth, *Samia cynthia* (Chan and Margoliash,

1966). Other species under study include the sheep, rabbit, and kangaroo (Margoliash and Smith, 1965). The cytochromes c of the cow and the pig are identical. That of wheat germ is under investigation (E. L. Smith, *unpublished*).

The results provided striking evidence for the evolution of all the living species in the series from a common ancestor (Margoliash, 1963).

It has long been known that the same complex molecules of many types are biologically synthesized by different living species. Examples of such molecules are found in the coenzymes, such as riboflavin and vitamin B_{12}, the amino acids, chlorophyll, and many other large and characteristic nonprotein molecules. Very few of these bear the imprint of evolutionary divergence.

The case of cytochromes is much more informative. The similarities and disparities between the cytochromes c of baker's yeast and of human beings are analogous to the situation one finds in comparing the French and Italian languages, whose common origin from the Latin tongue is indisputable. No one, even if he were unaware of history, would suggest that the similarities are an accidental result or are a coincidence brought about by the respective needs of two separate populations for communication among themselves. A comparison of either French or Italian with Japanese makes it obvious that languages do not tend to be similar when they originate in separate populations which are isolated from each other. The extent to which a language evolves within a country is variable. English, passed from father to son, bears the imprint of argot, neology, and other mutational stresses which change it with the passing of time. Latin is conserved, perhaps because it is carefully passed from Father to Father.

Yeast cytochrome c contains 65 amino acids that are identical with those occupying the corresponding positions in the sequence of human cytochrome c. On a random basis the chances are only 1 in 20 of the same amino acid occurring at two sites and only 1 in 20^{65} for coincidence occurring 65 times. Protein molecules, of course, are not random sequences of amino acids, and it is recognized that the occurrence of, for example, glutamic acid at position 4 in the cytochrome chain betokens a functional significance. In other cytochromes c, however, position 4 is occupied by lysine and alanine. Clearly, some flexibility exists at position 4.

In contrast, sites 70 through 80 are identical in all cytochromes c examined. This coincidence may indicate an absolute requirement for this sequence, or it may mean that, owing to chance and probability,

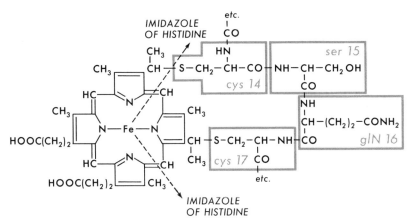

Fig. 6-1. The heme group and its attachment to human cytochrome c.

no evolutionary changes have become established in this region of the molecule, but this is unlikely in view of the fact that residues 70 to 80 are invariant in 15 different species (Margoliash and Smith, 1965; Smith, *unpublished*).

As pointed out by Margoliash (1963), there is a sequence not only in the cytochromes c but in other cytochromes that is concerned with the attachment of the heme group as shown in Fig. 6-1. This sequence is included in residues 13 to 19 and is shown for the peptide chains of various species in Table 6-1. The remainder of the sequence from *Chromatium* protein, corresponding to residues 20 to 36, is distinctly different from that of the human cytochrome c. The neighboring sequences in *Pseudomonas* cytochrome$_{551}$, corresponding to residues 3 to 12 and 20 to 24, also are distinctly different from the cytochromes

Other Proteins Showing Evolutionary Changes

c and from the related regions in *Chromatium* protein. Evidently the sequence cys. X. Y. cys. his is essential for the attachment of heme and so is completely conserved during evolution. The relation of the two cysteine residues to the attachment of the heme group in cytochrome *c* is shown in Fig. 6-1. Figure 6-1 shows also a suggested relationship to histidine residues 18 and 26; however, in one of the

Table 6-1

	13	14	15	16	17	18	19
Cytochrome *c* human	lys	cys	ser	glN	cys	his	thr
Cytochrome *c* horse	lys	cys	ala	glN	cys	his	thr
Cytochrome *c* yeast	arg	cys	glu	leu	cys	his	thr
Chromatium protein[a]	lys	cys	ser	glN	cys	his	thr
Cytochrome$_{551}$ *Pseudomonas*[b]	gly	cys	val	ala	cys	his	ala

[a] Dus, Bartsch, and Kamen (1962). [b] Ambler (1962, 1963).

cytochromes *c*, his 26 is replaced by glN (Heller and Smith, 1965). The peptide in *Chromatium* protein that binds heme (Dus et al., 1962) illustrates the phenomenon of partial gene duplication, followed by amino acid changes and the appearance of a two-residue gap. This peptide is related to the sequence 10 to 24 (Table 6-2) in the cytochromes *c* as shown below for a portion of tuna cytochrome *c* (in the first line). A deletion is postulated between residues 19 and 20.

... phe val glN lys cys ala glN cys his thr – val glu asN gly gly ...
NH_2-phe ala gly lys cys ser glN cys his thr leu val ala asp glu gly
                  ser ala lys cys his thr phe – – asp glu gly ser
Base changes 1 1   1   0 0 0   1   0   0   0

The homology between the cytochromes *c* is shown in Table 6-2 and in Figs. 6-2 and 6-3. The divergence of the various chains is correlated with the taxonomic separation of the five species that are listed. The phylogenetic significance of the comparison of the polypeptide chains has been discussed at length in illuminating articles by Margoliash (1963) and by Margoliash and Smith (1965).

The cytochromes *c* afford the best available example of the manner in which certain regions of a protein molecule have been conserved in a wide range of different species during hundreds of millions of years. There are 104 amino acids in the polypeptide chains of most of the cytochromes; the chain is extended by four to six amino acids at the NH$_2$-terminal end in the case of yeasts, *Neurospora*, and of a moth (*Samia cynthia*). Five of the cytochromes each have an amino acid deleted near the carboxy-terminal end.

The remainder of the molecules shows a number of similarities and differences. The similarities include the following:

1. Cys at 14 and 17, his at 18; involved in combination with the heme group. Margoliash and Smith have pointed out that this hemochrome area may be considered as extending from residues 11 to 33, but it is subject to variations in 10 of the 23 sites.

2. The following amino acids have been found to be invariant in all species studied: eight glycines at sites 1, 6, 29, 34, 41, 45, 77, and 84; six lysines; and from one to three of all the other amino acids except serine, as shown in Table 6-2. It has been pointed out that the gly residues, because of their lack of side chains, are good crossing-over points for the polypeptide chains during folding (E. L. Smith, *unpublished*).

3. An identical sequence of 11 amino acids in residues 70 to 80.

4. A notable lack of helical regions: no α helices in the first 80 residues and not more than 10 percent α helix in the entire chain as evidenced by amino acid sequences and optical rotatory dispersion (E. L. Smith, *unpublished*).

5. Considerable homology among hydrophobic sites, such as phe/leu/ilu or val/ileu (Margoliash, 1963). Such interchanges may be influenced by the facility with which base changes can take place in the coding triplets that are responsible for the changes.

On the other side of the coin, it is evident that many of the sites in the cytochrome are subject to great evolutionary flexibility. There are seven amino acid interchanges at residue 89, and these include one basic, two acidic, and three neutral amino acids. Four base changes at the coding sites of the residue, arranged in various combinations,

Table 6-2. Comparison of the Amino Acid Sequences of Cytochromes *c* from Human, Horse, Chicken, Tuna, and Baker's Yeast Sources and Various Other Sources; DNA Coding Triplets Corresponding to the Human Cytochrome *c* Sequence and Corresponding to Species Differences from Human Cytochrome *c*

Source of cytochromes *c*:		1								10	
Human	Acetyl-NH-gly	asp	val	glu	lys	gly	lys	lys	ilu	phe	ilu
Horse	Acetyl-NH-gly		ilu								val
Chicken	Acetyl-NH-gly		val	ala							val
Tuna	Acetyl-NH-gly	ser	ala	lys			ala	thr	thr		val
Yeast	NH₂-thr glu phe lys ala gly	asN	ser					asN	leu		lys
Others	gly val pro										thr
	NH₂-pro ala pro	glu glN									
		ser									

DNA coding triplets:											
Corresponding to human cytochrome *c* sequence		ACd	GAe	TTb	AAe	GCe	GGd	GAb	GTb	AAe	ATb	TTb ATA
Corresponding to species differences from human cytochrome *c*	{ CCd GCd CCb	GGb GTb CCe		AGb	ATb	GCe	AAb	GAe	GCe	ACe	ACb GTA
		GAe CAe		AAb	GCb					AAb	CTb AAA
		TCe		TCb							ACA

Source of cytochromes *c*:								20			
Human	met	lys	cys	ser	glN	cys	his	thr	val	glu	lys	his
Horse	glN			ala								
Chicken	glN			ser								
Tuna	glN			ala	leu						asN	pro
Yeast	thr		arg	glu	glu				gly	glu	ala	thr lys
Others											gly	

DNA coding triplets:											
Corresponding to human cytochrome *c* sequence	ATG	AAe	TGb	TCd	CAe	TGb	CAb	ACd	GTd	GAe	AAe	GGd	GGd AAe CAb
Corresponding to species differences from human cytochrome *c*	{ CAG AGe			GCd	CTe						AAb		CCe CTb
	ACG			GAe	GAe				GGd	GAe	GCd		ACe CAe
									GGd	AAb	CTd	

Source of cytochromes c:

				30										40	
Human	lys	thr	gly	pro	asN	leu	his	gly	leu	phe	gly	arg	lys	thr	gly	glN
Horse																
Chicken																
Tuna	val						try		ilu				his			
Yeast	val						his		phe			ser	his	ser		
Others	ilu				ala		asN				ser					ser

DNA coding triplets:

Corresponding to human cytochrome c sequence	AAe	ACd	GGd	CCd	AAb	CTb	CAb	GGd	CTb	TTb	GGb	CGd	AAe	ACb	GGd	CAe
									ATb	TAb			CAb	TCb		TCe
Corresponding to species differences from human cytochrome c	GTd						TGe		TTb							
	ATb				GCb		AAb					AGb				

Source of cytochromes c:

									50						
Human	ala	gly	tyr	ser	tyr	thr	ala	asN	lys	asN	lys	gly	ilu	ilu
Horse			phe	thr						asp				thr
Chicken			phe	ser						asp	ser			thr
Tuna			phe	ser						asp	lys	asN	val	val
Yeast	val	pro	tyr	ala	ser	asN	ala		ilu	asp	ala			leu
Others											glN	ala		glu
											arg			

DNA coding triplets:

Corresponding to human cytochrome c sequence	GCd	CCd	GGd	TAb	TCb	TAb	ACd	GCb	GCd	AAb	AAe	AAb	AAe	GGd	ATb	ATb
				TTb	ACb		TCd	GAb		ATA	AGb			AAb	GTb	ACb
Corresponding to species differences from human cytochrome c	GTd	GAe			GCb			AAb			AAe					GTb
	CAe										GCd					CTb
	GTd										CAe	GCe				GAe
	GAb										CGe					

Table 6-2. (*Continued*)

Source of cytochromes *c*:		60								70			
Human	try	gly	glu	asp	thr	leu	met	glu	tyr	leu	glu	asN	pro	lys	tyr
Horse		lys		glu											
Chicken		gly		asp											
Tuna		asN	asN	asp					asp			thr			
Yeast		asp	glu	asN	asN	met	ser	asp							
Others		ala		pro			leu	glN							
							phe								

DNA coding triplets:

Corresponding to human cytochrome *c* sequence	TGe	GGd	GAe	GAb	ACb	TTG	ATG	GAe	TAb	CTd	GAe	AAb	CCd	AAe	TAb
Corresponding to species differences from human cytochrome *c*	AAe	AAb	GAe	AAb		ATG	TCd	GAb					ACe		
	AAb		GAe	CCd			TTG	CAe							
	GAb						TTb								
	GCd														

Source of cytochromes *c*:					80								90	
Human	ilu	pro	gly	thr	lys	met	ilu	phe	val	gly	ilu	lys	lys	glu	glu
Horse									ala					thr	
Chicken									ala					ser	
Tuna									ala					gly	
Yeast							ala		gly		leu			lys	asp
Others						ser	val		thr					asp	asN
														ala	
														asp	

DNA coding triplets:

Corresponding to human cytochrome *c* sequence	TAb	CCd	GGd	ACd	AAe	ATG	ATb	TTb	GTd	GGd	ATb	AAe	AAe	GAe	GAe
Corresponding to species differences from human cytochrome *c*						GCe	GCd		GCe		CTb	AGb		GAe	ACd	GAb
						GTb	GGd		ACe					ACe	AGb	
							ACd							GCe	GGd	
														GAb	AAe	
															GAb	
															AAb	

Source of cytochromes c:									100					
Human	arg	ala	asp	leu	ilu	ala	tyr	leu	lys	ala	thr	asN	glu—COOH
Horse		glu								ala	thr	asN	glu—COOH
Chicken		val							asp	ala	thr	ser	lys—COOH
Tuna		glN		val				(del)		ser	thr	ala	ser—COOH
Yeast		asN		ilu				lys		ala	cys	(del)	glu—COOH
Others		thr			ilu	thr	phe	met	(del)	glu	ser	lys	ala
									leu				lys
DNA coding triplets:													
Corresponding to human cytochrome c sequence	CGd	GCd	GAb	CTd	ATb	GCd	TAb	TTG	AAe	GCd	ACb	AAb	GAe
Corresponding to species differences from human cytochrome c		GAe			ATb	ACd	TTb	ATG	GAb	TCb	TGb	AGb	AAe
		GTd						CTe	GAe	GAe	TCb	GCd	GCd
		CAe										AAe	
		ACd											

199

are sufficient to account for the seven amino acids. Other sites exhibit comparable versatility. The commonest interchange is ilu/val, which represents not only an alteration between two very similar hydrophobic amino acids but also a single-base transition between A and G in any of three pairs of coding triplets. There are various exceptions to any general rule about the persistence of basic residues, hydrophobic residues, proline, and glycine at specific sites, as may be seen in Table 6-2, which reveals interchangeability among glu, ala, and lys; ilu, val, and lys; lys and pro; pro, glu, glN, and val; and gly, lys, and asp and contains other examples of similar interchanges at various sites. Clearly, any conclusion about the conservation of specific residues should include a statistical examination of the possibility that such residues may have escaped change by the rules of chance. One simple test for this is to examine the Poisson distribution of changes, or "hits," in the loci of base sequences of the genes corresponding to the cytochrome c chains (Table 6-2). The Poisson distribution is an extension of the binomial theorem which is useful for the examination of randomness. Let us consider a sequence of 100 coding triplets in a gene. If base changes occur uniformly and by chance throughout the sequence, the probability that any one triplet will receive a single base change will increase steadily with the total number of such changes. As this number increases, the chances also increase that a triplet will receive two base changes. The Poisson formula states that after, for example, 60 hits on 100 sites, the average distribution will be unhit sites, 55; sites hit once, 33; twice, 10; three or more times, 2. The hits must be examined in the base sequences rather than the amino acid sequences, since the latter are secondary to the former. The results are shown in Table 6-3. The number of sites with more than one change is probably too small to be reliable, but the suggestion of possible randomness is interesting.

How many of the 39 presently "invariant" amino acid residues in the cytochromes c may be found to be subject to change when enough species are examined? The sequence from 70 to 80 is inviolate so far, and it seems unlikely that this is due to chance. On the other hand, evolutionary changes in the cytochromes c are evidently extremely slow.

Results with other species, especially those having rapid reproductive cycles such as yeasts and molds, will probably indicate that some of the presently invariant residues are actually variable. Indeed, Narita and Titani (1965) indicate that residues 37, 66, and 100, formerly constant, have been replaced in *Candida krusei*. This has reduced the number of invariant residues to 36.

Table 6-3. Distribution of Base Changes in Coding Triplets in Table 6-2 Compared with Poisson Distribution

No. of Base Changes	Percent of Sites Corresponding	Expected from Poisson Distribution If 64 Percent of Sites Are Unchanged
0	63.8	(64)
1	30.7	29
2	5.5	6
3	None	1

The polypeptide chains of the various cytochromes were written in terms of the coding triplets (Table 6-2) with the reservations that (1) in most cases the third base cannot be specified without ambiguity and (2) arg, leu, and ser each has two sets of codes, and it is often not possible to guess which of the two is most likely to occupy a given site.

Perhaps the most interesting observation that arises from this speculative attempt to infer the base sequences in the cytochrome genes is that the number of base changes necessary to provide so many amino acid changes is comparatively small. The base sequences present a more closely related picture than the amino acid sequences (Figs. 6-2 and 6-3). Part of the reason is that there are 20 different amino acids and only 4 different bases, but it is also true that the majority of the amino acid differences between the homologous residues in the various cytochromes *c* may be written in terms of single-base changes in the coding triplets. As might be expected, the number of changes involving the alteration of two bases in a coding triplet tends to increase as the number of differences between two cytochromes *c* increases and as the taxonomic relationship between two species becomes more remote (Table 6-4).

202 Other Proteins Showing Evolutionary Changes

Calculated in terms of estimated base changes, a linear rate of such changes being assumed, the evolutionary time relationships between the cytochromes c may be roughly estimated as shown in Table 6-4. This procedure is based on each species having the same mutation

Table 6-4. Base Changes in the Evolutionary Separation of the Cytochromes c

Comparison	Coding Triplet Changes			Base Changes		
	1-Base	2-Base	3-Base	Transitions	Transversions	Total
Human/horse	9	3	0	6	9	15
Human/chicken	9	3	1	8	10	18
Average	...	...	...	...	...	16.5
Human/tuna	13	8	1	15	16	32
Horse/tuna	10	6	1	11	14	25
Chicken/tuna	12	7	1	17	12	29
Average	...	...	...	...	...	28.1
Human/yeast	24	11	1	21	28	49
Horse/yeast	23	15	1	22	34	56
Chicken/yeast	22	17	1	24	35	59
Tuna/yeast	25	18	1	30	34	64
Average	...	...	...	...	...	57.0
Human/*Neurospora*	31	14	0	24	35	59
Horse/*Neurospora*	26	17	0	23	37	60
Chicken/*Neurospora*	26	15	0	22	34	56
Tuna/*Neurospora*	25	21	1	31	39	70
Yeast/*Neurospora*	27	14	0	20	35	55
Average	...	...	...	...	...	60.0
Total[a]	147	88	...	52	78	130

[a] All species in Table 6-2.

rate, which is another way of saying that the DNA molecules in all species must all change at the same rate. This in itself is doubtful for various reasons, but there is another complication: evolution must proceed by natural selection, and this will take place much more rapidly in a species with a short reproductive cycle, such as yeast, which is measured in hours, than in the large vertebrates, which may take

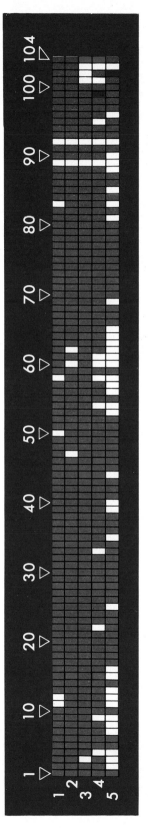

Fig. 6-2. Amino-acid replacements in possible coding sequences for cytochromes c: (1) human, (2) horse, (3) chicken, (4) tuna, (5) yeast. White areas indicate amino acid differences from (1); green areas indicate differences from (1) but identity of other chains with each other; dark areas indicate gaps.

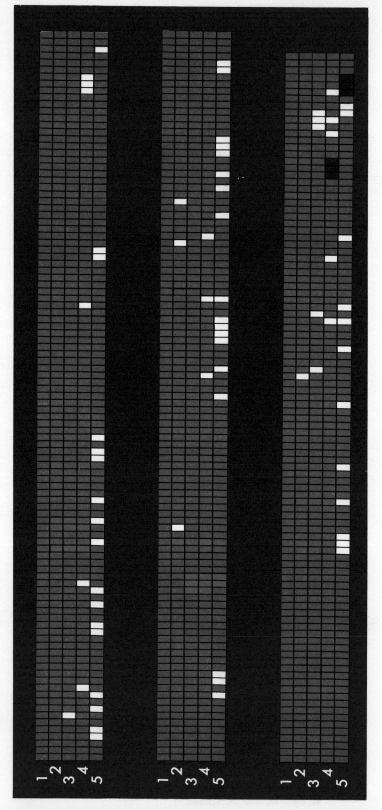

Fig. 6-3. Base replacements in possible coding sequences for cytochromes c: (1) human, (2) horse, (3) chicken, (4) tuna, (5) yeast. White areas indicate base differences from (1); green areas indicate differences from (1) but identity of other chains with each other; dark areas indicate gaps.

a number of years to come to sexual maturity. Clearly, one would expect the cytochrome c of yeast to have developed much further from its ancestral archetype than have the vertebrate cytochromes c. It is not possible to express this idea quantitatively, but it perhaps renders doubly impressive the similarity shown between yeast cytochrome c and those of the other species in Table 6-2. The yeast/vertebrate difference of 0.183 base change per site and the *Neurospora*/vertebrate difference of 0.195 base change per site are about twice as great as the tuna/warm blooded species difference of 0.090 base change per site. The *Neurospora*/yeast/vertebrate comparison hints that the *Neurospora* and yeast may have diverged more rapidly from each other than the warm-blooded animals from the tuna. It also suggests that the common archetype of all the cytochromes c in Table 6-4 may have resembled those of the present vertebrates more closely than those of the yeasts and molds. This would be in keeping with the more rapid reproductive rate of microorganisms, as noted by Heller and Smith (1965).

A rough estimate of the number of invariable amino acid residues in the cytochromes c is about 34. On the basis of the Poisson distribution, for the remaining 70 variable residues and 30 unchanged sites, the yeast/vertebrate comparison is as follows:

	Calculated	*Observed*
Single-base changes	25	23.5
Two or more base changes	15	16.2

The conclusion is drawn that the changes in the variable sites have occurred essentially at random, although the molecule is obviously subject to a considerable degree of conservative restraint as shown by the persistence of many residues in all species (Margoliash, 1963; Margoliash and Smith, 1965). The evolutionary process is evidently still in progress in the cytochrome c molecules. It proceeds more slowly in the cytochromes c than in the globins.

The juxtaposition of the base sequences is visualized in Fig. 6-3, which shows the relative paucity of changes and the close similarity of the DNA in different species.

Other Proteins Showing Evolutionary Changes

FIBRINOGEN. A comparison of the amino acid sequences in fibrinopeptides A and B of seven different mammals was made by Doolittle and Blombäck (1964). These two peptides are released from fibrinogen by the action of thrombin, a protease that splits four specific arg-gly bonds in fibrinogen with the following results:

Fibrinogen $\xrightarrow{\text{Thrombin}}$ 2 A-peptides + 2 B-peptides + fibrin monomer

Fibrin monomers $\longrightarrow$ polymerization $\longrightarrow$ clot

The primary structure of the A and B fibrinopeptides from various species is shown in Table 6-5, adapted from the article by Doolittle and Blombäck (1964). The A and B peptides evidently both start from the NH_2-terminal end of longer polypeptide chains which are broken by fibrin at the arg-COOH residues shown in Table 6-5. The base changes per 100 sites in terms of species differences, calculated in a manner similar to that shown in Table 6-2 for the cytochromes c, are in Table 6-6. There is a slight correlation between taxonomy and divergence in terms of base changes, but the agreement is only very rough; for example, the cloven-hoofed animals seem to be more closely related to each other than they are to human beings or rabbits. The polypeptide sequences in Table 6-5 are too short to permit a quantitative comparison of closely related species of animals. This is evident from the fact that the sequences are identical for two different species, the sheep and the goat, the proteins of which undoubtedly have many differences. From this it follows that the differences shown in Table 6-6 between cattle and sheep and between cattle and reindeer probably cannot be expected to reveal more than a rough difference between each pair of species and do not suffice to indicate whether it is the reindeer or the sheep that is more closely related to cattle. This is brought out further by the following: Let us assume that the three species are descended from a common ancestor and that sheep branched off subsequently from the later ancestor of sheep and reindeer. Then the sheep/reindeer difference should be less than the sheep/cattle difference, which is the case. However, the cattle/sheep difference should then be

Table 6-5. Amino Acid Sequences of Fibrinopeptides A and B from Various Mammals

Fibrinopeptides A:

Cattle	NH$_2$-glu	asp	gly	ser	asp	pro	ser	gly	asp	phe	leu	thr	glu	gly	gly	gly	val	arg	COOH
Sheep and goat	NH$_2$-ala	asp	asp	ser	asp	pro	val	gly	glu	phe	leu	ala	glu	gly	gly	gly	val	arg	COOH
Reindeer	NH$_2$-ala	asp	gly	ser	asp	pro	ala	gly	glu	phe	(leu	ala	glu	gly	gly	gly	val)	arg	COOH
Pig	NH$_2$-ala	glu	val	glN	asp			lys	glu	phe	leu	ala	glu	gly	gly	gly	val	arg	COOH
Human	NH$_2$-ala	asp	ser	gly	glu			gly	asp	phe	leu	ala	glu	gly	gly	gly	val	arg	COOH
Rabbit	NH$_2$-val	asp	pro	gly	glu			thr	ser	phe	leu	(thr	glu	gly	gly	gly)asp	ala	arg	COOH

Fibrinopeptides B:

Cattle	NH$_2$-glN	phe	pro	thr	asp	tyr	asp	glu	glN	asp	asp	arg	pro	lys	val	gly	leu	gly	ala	arg COOH
Sheep and goat	NH$_2$-gly	tyr	leu		asp	tyr	asp	glu	val	asp	asp	asN arg	ala	lys	leu	pro	leu	asp	ala	arg COOH
Reindeer	NH$_2$-glN	leu	ala		asp	tyr	asp	glu	val	(glu	asp	his) arg	ala	lys	leu	his	leu	asp	ala	arg COOH
Pig	NH$_2$-ala	ilu			asp	tyr	asp	glu	glu	asp	asp	gly arg	pro	lys	val	his	val	asp	ala	arg COOH
Human	NH$_2$-glN	gly	val	asN	asp						asN glu					glu	gly	phe phe ser	ala	arg COOH
Rabbit	NH$_2$-ala	asp			asp	tyr	asp	glu								asp	val	pro leu asp	ala	arg COOH

Source: Adapted from Doolittle and Blombäck (1964).

equal to the cattle/reindeer difference, which is not so. The disparity is not disturbing when viewed statistically; one should expect to take an adequately large sample of DNA from each of several closely related species before expecting to make a taxonomically significant comparison. In spite of this, a fibrinopeptide sequence of only nine amino acids, corresponding to 27 DNA base pairs, was selected by Doolittle and Blombäck (1964) and by Epstein and Motulsky (1965) for making evolutionary comparisons among five hoofed mammals. This procedure

Table 6-6. Comparison of Base Changes Needed to Produce the Amino Acid Substitutions in the Fibrinopeptides Shown in Table 6-5, Estimated by Methods Similar to Those used for the Cytochromes c. (See Tables 6-2 and 6-4.)

Fibrinopeptides	Percent Base Changes from				
	Cattle	Sheep/Goat	Reindeer	Pig	Human
Sheep and goat	17	...	...	...	...
Reindeer	13	9	...	...	...
Pig	17	16	15	...	...
Human	19	21	18	21	...
Rabbit	23	20	22	20	23

was obviously inaccurate in that it showed less difference between cattle and pigs than between cattle and sheep. Even in the case of the cytochromes c, the sample, which represents about 312 DNA base pairs, is too small for making taxonomic comparisons of mammalian species; for example, the cytochromes c of cattle and the pig are identical, while those of cattle and the horse show differences that correspond to four base changes.

In the article by Doolittle and Blombäck (1964), the polypeptide chains were aligned without "gaps." This procedure assumes that the differences in length between the polypeptide chains of the fibrinogens in the various species are due to shortening of the gene, taking place at the end corresponding to the initiation of mRNA synthesis. This seems unlikely for several reasons: it is probable that a specific initiation site on the DNA molecule exists for starting transcription; evidence is accumulating for a specific initiator sequence for mRNA translation; internal deletion in a gene by unequal crossing-over seems to be

Other Proteins Showing Evolutionary Changes 209

indicated as a frequent evolutionary occurrence by examples in the globins, chymotrypsinogens, and trypsinogen, described elsewhere in this chapter; and, finally, the homology of the amino acid residues in the fibrinopeptides of different species is improved by assuming the existence of gaps produced by internal deletions. This is shown in Table 6-5.

The N-terminal amino acid in the fibrinopeptides-B of cattle, reindeer, and human beings was assumed to be glutamine by Doolittle and Blombäck because of their finding of a pyrrolidone ring in this position. Apparently an N-terminal glutamine can condense into such a ring. With the exception of glutamine, the amino acids in the NH_2-terminal position of fibrinopeptides A and B have coding triplets that start with a purine nucleotide. This is true also for the globin polypeptide chains (Table 5-2), the cytochromes c (Table 6-2), except that of *Candida krusei*, the haptoglobins (p. 223), and ferredoxin (p. 228). Perhaps this is related to the fact that the RNA polymerase reaction initiates RNA molecules with purines at the starting point.

TRYPSINOGEN AND CHYMOTRYPSINOGEN. The pancreas of animals secretes a slightly alkaline juice containing the precursors of several proteolytic enzymes, including trypsinogen and chymotrypsinogen. There are two different chymotrypsinogens, A and B, in bovine pancreatic juice but only one in that of the pig or the dog. Knowledge of the primary structure of bovine chymotrypsinogen B is incomplete. Only one trypsinogen occurs in the pancreatic juice, but differences are known to occur between the trypsinogens of different species. These have not been explored in terms of amino acid sequences. Trypsinogen is enzymatically inactive but is converted to the active trypsin by the removal of an amino-terminal peptide, val-(asp)$_4$-lys, by the action of enterokinase or of trypsin itself. The activation of chymotrypsinogen A takes place in several steps, which are initiated by the action of trypsin in splitting an arg-ilu bond. Other bonds are then broken by chymotrypsin, liberating the dipeptides ser-arg and thr-asN to leave the active form of the enzyme, α chymotrypsin. This contains three peptide chains; the A chain of 13 residues, the B chain of 131 residues, and the C chain of 98 residues, which contains the active center (residues 193 to 198). The A, B, and C chains are connected by two disulfide bridges. There are three other disulfide bridges, one is internal in the B chain and two are internal in the C chain.

Trypsin splits polypeptide chains primarily at points between the carboxyl groups of arg and lys and the next amino acid. The action of chymotrypsin is less specific. Its sites of action are at the carboxyl groups of phe, try, and tyr peptide linkages, but it also attacks the bonds connecting asN, his, leu, and met to the next amino acid.

The complete amino acid sequences of bovine chymotrypsinogen A and trypsinogen have been discovered with the exception of a tetrapeptide occupying residues 84 to 87 in trypsinogen (Hartley, 1964; Walsh and Neurath, 1964; Hartley et al., 1965). The two sequences are shown in alignment in Table 6-7, and it is evident that again we are confronted with an example of gene duplication and subsequent evolution.

The active site in both enzymes is at the serine residue in position 195 in the chymotrypsinogen A chain and in 185 in trypsinogen (198 in Table 6-7). The relationship of the two chains as postulated in Table 6-7 reveals the presence of four gaps in chymotrypsinogen A and six in trypsinogen. The first nine amino acids in chymotrypsinogen A are missing from trypsinogen, perhaps because of either deletion, crossing-over, or partial rather than complete gene duplication.

Hartley et al. (1965) have determined the primary structures of portions of the molecules of bovine chymotrypsinogen B and another pancreatic enzyme, porcine elastase. These are included in Table 6-7, and the information is sufficient to indicate that these two enzymes are products of additional duplication of the same archetypal gene that gave rise to the present genes of chymotrypsinogen A and trypsinogen. The alignment of residues and assignment of gaps in Table 6-7 follow for the most part the suggestions made by Hartley et al. (1965).

A comparison of the two complete sequences of chymotrypsinogen A and trypsinogen in Table 6-7 shows the relationships of Table 6-8 in terms of coding triplets.

Table 6-7. Complete Amino Acid Sequences of Bovine Chymotrypsinogen A, Trypsinogen (Lines 2 and 3) Together with Partial Sequences of Bovine Chymotrypsinogen B and Porcine Elastase (Lines 4 and 5)

	1	2	3	4	5	6	7	8	9	10	11	12	13	14	15	16	17	18	19	20	21	22	23
Chymotrypsinogen A	cys-	gly-	val-	pro-	ala-	ilu-		glN-	pro-	val-	leu-	ser-	gly-	leu-	ser-	arg-	ilu-	val-	asp-	glu-	glu-	ala-	val-
Trypsinogen											val-	asp-	asp-	asp-	lys-	ilu-	val-	gly-	gly-	tyr-	thr-	cys-	gly-
Chymotrypsinogen B	cys-	gly-	val-	pro-	ala-	ilu-		glN							ala-	arg-	ilu-	val-	gly-	asx-	asx-	glx	
Elastase																							

	24	25	26	27	28	29	30	31	32	33	34	35	36	37	38	39	40	41	42	43	44	45	46
Chymotrypsinogen A	pro-	gly-	ser-	try-	pro-	try-	glN-	val-	ser-	leu-	glN-	asp-	lys-	thr-	gly-	phe-	his-	phe-	cys-	gly-	gly-	ser-	leu-
Trypsinogen	ala-	asN-	thr-	val-	pro-	tyr-	glN-	val-	ser-	leu-	asN-	0-	0-	ser-	gly-	tyr-	his-	phe-	cys-	gly-	gly-	ser-	leu-
Chymotrypsinogen B																	his-	phe-	cys-	gly-	gly-	ser-	leu
Elastase																try-	ala-	his-	thr-	cys-	gly-	gly-	thr-	leu

	47	48	49	50	51	52	53	54	55	56	57	58	59	60	61	62	63	64	65	66	67	68	69
Chymotrypsinogen A	ilu-	asN-	glu-	asN-	try-	val-	thr-	ala-	ala-	his-	cys-	gly-	val-	thr-	thr-	ser-	asp-	val-	val-	ala-	gly-
Trypsinogen	ilu-	asN-	ser-	glN-	try-	val-	val-	ser-	ala-	ala-	his-	cys-	tyr-	lys-	ser-	gly-	ilu-	glN-	val-	arg-	leu-	0-	gly-
Chymotrypsinogen B									ala-	ala-	his-	cys-	gly-	val-	thr-	thr-	ser-	asp-					
Elastase									ala-	ala-	his-	cys-	val-	asp-	val-	asp-	arg-	glx					

	70	71	72	73	74	75	76	77	78	79	80	81	82	83	84	85	86	87	88	89	90	91	92
Chymotrypsinogen A	glu-	phe-	asp-	glN-	gly-	ser-	ser-	ser-	glu-	lys-	ilu-	0-	glN-	lys-	leu-	lys-	ilu-	ala-	lys-	val-	phe-	lys-	asN-
Trypsinogen	glu-	0-	asp-	asN-	ilu-	asN-	val-	val-	glu-	gly-	asp-	glu-	glN-	phe-	ilu-	ser-	ala-	ser-	lys-	ilu-	val-	his-	
Chymotrypsinogen B																							
Elastase																							

Table 6-7 (*continued*)

```
                     93   94   95   96   97   98   99  100  101  102  103  104  105  106  107  108  109  110  111  112  113  114  115
Chymotrypsinogen A   ser- lys- tyr- asN- ser- leu- thr- ilu- asN-asN-asN-asN- ilu- thr- leu- leu- lys- leu- ser- thr- ala- ala- ser- phe
Trypsinogen          pro- ser- tyr- asN(pro, leu, thr, asN) asN-asN-asp- ilu- met-leu- ilu- lys- ser- ala- ala- ser- leu
Chymotrypsinogen B
Elastase

                    116  117  118  119  120  121  122  123  124  125  126  127  128  129  130  131  132  133  134  135  136  137  138
Chymotrypsinogen A   ser- glN-thr- val- ser- ala- val- cys- leu- pro- ser- ala- ser- asp- asp- phe-ala- ala- gly- thr- cys  val
Trypsinogen          asN-ser- arg- val- ala- ser- ilu- ser- leu- pro thr- ser- cys-  0-   0- ala- ser- ala- gly- thr- glN-cys- leu-
Chymotrypsinogen B                                                                                                          leu- cys- ala-
Elastase                                   ala- val- cys- leu- pro- ser- ala- asp                  ala- asN-asN-ser- pro- cys- tyr

                    139  140  141  142  143  144  145  146  147  148  149  150  151  152  153  154  155  156  157  158  159  160  161
Chymotrypsinogen A   thr- gly- gly- try- gly- leu- thr- arg- tyr- thr- asN-ala- asN-thr- pro-asp- arg- leu- glN-glN-ala- ser- leu-
Trypsinogen          ilu- ser- gly- try- gly- asN-thr- lys- ser- ser- gly- thr- ser- tyr- pro-asp- val- leu- lys- cys- leu- lys- ala-
Chymotrypsinogen B   thr- thr- gly
Elastase

                    162  163  164  165  166  167  168  169  170  171  172  173  174  175  176  177  178  179  180  181  182  183  184
Chymotrypsinogen A   pro- leu- ser- asN-thr- asN-cys- lys- lys- tyr- try- gly- thr- lys- ilu- lys- asp- ala- met-ilu- cys- ala-
Trypsinogen          pro- ilu- leu- ser- asp- ser- cys- lys- ser- ala- tyr- pro- gly- glN-ilu- thr- ser- asN-met-phe-cys- ala-
Chymotrypsinogen B                   asN-thr- asp- cys- arg                                                          met-ilu- cys- ala-
Elastase                             ala- ilu- cys- ser- ser- ser- ser- tyr                                         met-val- cys- ala-
```

	185	186	187	188	189	190	191	192	193	194	195	196	197	198	199	200	201	202	203	204	205	206	207
Chymotrypsinogen A	gly-	0-	ala-	ser-	gly-	val-	0-	ser-	ser-	cys-	met-	gly-	asp-	ser-	gly-	gly-	pro-	leu-	val-	cys-	lys-	lys-	asN-
Trypsinogen	gly-	tyr-	leu-	glu-	gly-	lys-	gly-	lys-	asN-	ser-	cys-	glN-	gly-	asp-	ser-	gly-	gly-	pro-	val-	cys-	ser-	gly-	lys-
Chymotrypsinogen B	gly-	0-	gly-	asp-	gly-	val		ser-	cys-	met-	gly-	asp-	ser-	gly-	gly-	pro-	leu-	val-	cys-	glN-	lys
Elastase	gly						arg-	ser-	gly-	cys-	glN-	gly-	asp-	ser-	gly-	gly-	pro-	leu-	his-	cys-	leu-	val-	asN-

	208	209	210	211	212	213	214	215	216	217	218	219	220	221	222	223	224	225	226	227	228	229	230
Chymotrypsinogen A	gly-	ala-	try-	thr-	leu-	val-	gly-	ilu-	val-	ser-	try-	gly-	ser-	ser-	thr-	cys-	ser -	thr-	0-	pro-	gly-
Trypsinogen					leu-	glN-	gly-	ilu-	val-	ser-	try-	gly-	ser-	0-	gly	cys-	ala	-glN-	lys-	asN-lys-	pro-	gly-
Chymotrypsinogen B																			cys-	ser -	thr-
Elastase	glN-tyr									val-	ser-	arg-	leu-	gly-	cys-	asN-val-	thr-	arg-	lys-	pro-	thr

	231	232	233	234	235	236	237	238	239	240	241	242	243	244	245	246	247	248	249
Chymotrypsinogen A	val-	tyr-	ala-	arg-	val-	thr-	ala-	leu-	val-	asN-try-	val-	glN-glN-thr-	leu-	ala-	ala-	asN-COOH
Trypsinogen	val-	tyr-	thr-	lys-	val-	cys-	asN-tyr-	val-	ser-	try-	ilu-	lys-	glN-	thr-ilu-	ala-	ser-	asN-COOH
Chymotrypsinogen B	val-	phe-																	

Asx = either asp or asN; glx = either glu or glN

The numbering corresponds to the total number of sites, including gaps (shown by 0), as postulated for chymotrypsinogen A and trypsinogen. Gaps have been placed in terms of the minimum number of base changes at homologous loci.

Other Proteins Showing Evolutionary Changes

The number of amino acid changes corresponds to 0.55 per residue. The total number of base changes, 186, corresponds to 0.27 change per locus for the 678 base pairs compared in the two genes. This is almost identical with the comparison between α hemoglobin and β or γ hemoglobins. The ratio of transitions to transversions is 1:1.8, also

Table 6-8

	No.	Percent
Loci compared	226	100
Identical amino acid residues	102	45
Single-base changes	64	28
Two-base changes	58	26
Three-base changes	2	0.9
Base changes	186	...
Transitions	69	...
Transversions	117	...

almost identical with the ratio found when α hemoglobin is compared with β and γ hemoglobins.

Whether or not the duplication of the chymotrypsinogen A gene (CTR), which may be older than the trypsinogen gene (TR), owing to its greater length, took place at about the same time as the duplication of the α hemoglobin gene is impossible to conjecture, since nothing is known of the rate of evolutionary change of CTR and TR.

If the proportion of single changes to changes in two or more bases per triplet is distributed randomly in the CTR:TR comparison, then it is possible by means of the Poisson distribution to make a rough calculation of the number of "variable" loci which are potentially variable but as yet are unchanged. This proportion is 64:60, or about 36:34, giving a Poisson distribution ($m = 1.2$) of unchanged $= 30$, one change $= 36$, two or more $= 34$, so that the number of variable unchanged loci should be approximately 53 as compared with the 92 unchanged amino acid residues. Hence, about 39 of the amino acids should be invariable. There are three long identical sequences in the two molecules, corresponding in Table 6-7 to 40 to 48, 196 to 201, and 214 to 220, for a total of 22 residues. The latter sequence contains the active center of the enzymes. In addition, several of the cysteines are presumably invariable in view of their participation in disulfide bridges.

Only one cysteine is present in the long identical sequences mentioned above. Six proline residues occupy identical positions in the two chains. The suggestion that about 39 of the amino acids should be invariable does not seem to be unreasonable. Chymotrypsinogen A and trypsinogen appear as examples of an evolutionary step caused by gene duplication followed by a few deletions and many random base changes and marked by conservation of function through the preservation of essential sites. The end result for a group of animals is an increase in specialization and sophistication because of the acquisition of a new proteolytic enzyme and perhaps also because of the refinement of an older one. Both chymotrypsin and trypsin are well adapted to digesting the proteins of the food of their possessors who subsist by eating plants or other animals.

We can contemplate, in terms of amino acid substitutions and base changes, the ancient imprint of the aeons marked by randomly mutational events which forged the difference between these two molecules during several hundred million years, perhaps at a rate of about one base-pair change in 2 or 3 million years.

TOBACCO MOSAIC VIRUS. A comparison of the amino acid sequences in the coat proteins of *vulgare* and *dahlemense* strains of TMV was made by Wittmann and Wittmann-Liebold (1963). The results are summarized in Table 6-9 and are compared with the replacements in mutants. Fifty-eight of the 158 sites have been found to be subject to change. The longer unchanged sequences are 108 to 121, 87 to 96, and 34 to 42; perhaps these three are invariant regions, as has been suggested by Wittmann for the sequence 108 to 121.

A comparison of the *vulgare* and *dahlemense* sequences shows essentially the same features in these RNA viruses as were encountered in comparisons of the cytochromes and hemoglobins except for the absence of deletions. There are 23 single-base changes and 6 two-base changes in the coding triplets. If these follow the Poisson distribution, it might imply that there are about 60 invariant sites in the TMV protein polypeptide sequence of 158 residues. The ratio of transitions to transversions is narrower than in the other proteins that have been

216 Other Proteins Showing Evolutionary Changes

Table 6-9. Amino Acid Sequence of TMV Protein Strain *Vulgare* Compared with That of Strain *Dahlemense* and Mutants

Residue No.:										10		
Vulgare	AcNH	Ser	tyr	ser	ilu	thr	thr	pro	ser	glN	phe	val	phe
Dahlemense							ser						
Mutants {					ilu						leu	met	

Residue No.:							20					
Vulgare	leu	ser	ser	ala	try	ala	asp	pro	ilu	glu	leu	ilu	asN
Dahlemense				val								leu	
Mutants {			leu				ala	leu	thr	val			ser
							val	thr	val				

Residue No.:					30							
Vulgare	leu	cys	thr	asN	ala	leu	gly	asN	glN	phe	glN	thr	glN
Dahlemense	val			ser	ser								
Mutants {			ilu					ser					
			ala					lys					

Residue No.:		40							50		
Vulgare	glN	ala	arg	thr	val	val	glN	arg	glN	phe	ser	glN	val
Dahlemense					thr			glN				glu	
Mutants {								gly					
								lys					

Residue No.:						60					
Vulgare	try	lys	pro	ser	pro	glN	val	thr	val	arg	phe	pro	asp
Dahlemense				phe			ser						gly
Mutants {				leu			ala	ilu		gly		ser	

Residue No.:										
Vulgare	ser	asp	phe	lys	val	tyr	arg	tyr	asN	ala	val	leu	asp
Dahlemense	asp	val	tyr										
Mutants {	gly	gly							ser				

Residue No.:			80								90
Vulgare	pro	leu	val	thr	ala	leu	leu	gly	ala	phe	asp	thr	arg
Dahlemense			ilu						thr				
Mutants {				ala									

Table 6-9 (continued)

Residue No.:								100				
Vulgare	asN	arg	ilu	ilu	glu	val	glu	asN	glN	ala	asN	pro	thr
Dahlemense										glN	ser		
Mutants						gly			arg				

Residue No.:					110							
Vulgare	thr	ala	glu	thr	leu	asp	ala	thr	arg	arg	val	asp	asp
Dahlemense													
Mutants				met									

Residue No.:			120								
Vulgare	ala	thr	val	ala	ilu	arg	ser	ala	ilu	asN	asN	leu	ilu
Dahlemense									val				val
Mutants						gly			val	asp			thr
										ser			val

Residue No.:	130									140		
Vulgare	val	glu	leu	ilu	arg	gly	thr	gly	ser	tyr	asN	arg	ser
Dahlemense	asN			val					leu			glN	asN
Mutants					gly		ilu		phe	cys	lys		

Residue No.:						150					
Vulgare	ser	phe	glu	ser	ser	ser	gly	leu	val	try	thr	ser	gly
Dahlemense	thr				met								ala
Mutants					phe						ilu		

Residue No.:			158	
Vulgare		pro	ala	thr
Dahlemense				ser
Mutants		leu		

discussed. The totals are 21 transitions and 15 transversions in the *vulgare/dahlemense* comparison. Transitions predominate in the mutations, as would be anticipated from the fact that most of them were brought about by the action of nitrous acid.

We thus see the same type of mutational trend, a randomly scattered series of single-base changes, in the evolutionary divergence of two strains of a plant RNA virus as we encounter in the genes of higher organisms. The forces of natural selection doubtless are similar; TMV needs a coat protein to encase its strand of RNA by means of a helical sheath of polypeptide units, each containing 158 amino acids. The perpetuation of the virus will depend upon the conformation of its protein coat, which can vary within certain limits. It is known that the protein coat differs slightly in its helical winding in *dahlemense* as compared with *vulgare* (Caspar, 1963). In the former strain the turns of the helix are brought alternately closer together and further apart; in the latter strain the spacing is uniform. There is very little difference, however, in the conformation of the individual subunits of the two strains (Caspar, 1963). The survival of the virus will also depend upon the biological synthesis of its other proteins, which presumably include one or two RNA synthetases and a lysozymelike enzyme that will enable it to emerge from the cells of the host plant. Nothing is known of the nature of these other proteins, but it is obvious that the length of the TMV RNA molecule (about 6,000–7,000 bases) is sufficient to provide for the coding of several other proteins in addition to that of the coat.

INSULIN. Insulin was the first protein whose primary structure was determined (p. 18). It contains two polypeptide chains held together by two S-S bridges of cysteine residues. The A chain has 21 amino acid residues and the B chain has 30. Insulin was also the first protein in which species differences were located with respect to amino acids, in this case at A8, 9, and 10. Other substitutions were found later, and these are summarized in Table 6-10. As might be expected, the insulin of a fish, the bonito, shows greater differences from the mammalian insulins than are found by comparison of different mammals (Kotaki, 1963).

When the results in Table 6-10 are summarized, the differences known at present between various insulins are: unchanged sites, 30; single-base changes, 15; two-base changes, 6. The corresponding Poisson distribution $m = 0.528$ is 30, 16, 5, in good agreement with this small sample. In spite of this, it seems reasonably likely that a portion of the insulin primary structure is invariable; certainly, the cysteines at

Table 6-10. Amino Acid Substitutions in Various Insulins as Compared with Bovine Insulin

Site Number and Amino Acid	Amino Acids in Other Species		Site Number and Amino Acid	Amino Acids in Other Species	
A3-val	glu	B-II	B1-phe	ala	B-II
A4-glu	asp	rat	B2-val	ala	B-II
A5-glN	his	B-II[a]	B3-asN	lys	rat
A8-ala	thr	various	B4-glN	pro	B-II
A8-ala	pro	B-II[a]	B27-thr	glN	B-II
A9-ser	gly	SH	B30-ala	thr	hu
A9-ser	his	B-II[a]	B30-ala	ser	R
A10-val	ilu	various	B30-ala	gap	B-II
A10-val	thr	S-W			
A10-val	lys	B-II[a]			
A12-ser	asp	B-II[a]			

S = sheep, H = horse, S-W = sei whale, B-II = bonito II, hu = human, R = rabbit.

[a] Sequence of B-II not precisely established, but most probable interchanges are listed.

A7, B7, A19, B19 must be essential. There are 12 transitions and 15 transversions in the evolutionary changes, equivalent to 0.18 base change per site.

As mentioned on p. 19, the A and B chains of the insulins evidently originated from a single primitive gene that underwent partial duplication followed by evolutionary base changes and four deletions (Fig. 2-3). The base changes are transitions, 6; transversions, 7; average per site, 0.25, as contrasted with 0.09 base change per site for the beef/bonito comparison, from which it might be concluded that the insulin molecule antedates considerably the separation between mammals and fishes. Deletions may be the result of unequal crossing-over.

220 Other Proteins Showing Evolutionary Changes

Table 6-11

	25								33
Human	asp	ala	gly	glu	asp	glN	ser	ala	glu
Beef	asp	gly	glu	ala	glu	asp	ser	ala	glN
Sheep	ala	gly	glu	asp	asp	glu	ala	ser	glN
Pig	asp	gly	ala	glu	asp	glN	leu	ala	glu

CORTICOTROPINS. The adrenocorticotropic hormone of the pituitary consists of a polypeptide chain of 39 amino acid residues. Of these, 1 to 24 and 34 to 39 are constant in human, beef, pig, and sheep corticotropins. There is a variable sequence from 25 to 33 (Table 6-11), which may be written in terms of coding triplets (Table 6-12). The interplay of amino acids between these sequences may thus need only 10 base changes in the 27 sites in the four chains of DNA.

Table 6-12

Human	GAb GCd GGe GAe GAb CAe TCe GCd GAe
Beef	GAb GGd GAe GCd GAe GAb TCe GCd CAe
Sheep	GCb GGd GAe GAb GAb GAe GCe TCd CAe
Pig	GAb GGd GCe GAb GAb CAe TTe GCd GAe

A further relationship in this group of polypeptides is found between the first 10 residues of corticotropin and the α and β melanocyte-stimulating hormones (MSH), in Table 6-13. The genes for the MSH molecules were perhaps formed by partial duplications of a portion of a gene which included the information for corticotropin. This was followed by a few amino acid changes at variable sites. The same gene evidently had a common origin with that of a portion of the gene containing the information for the lipotropic hormone of the pituitary (Li et al., 1965).

Crossing-Over: The Haptoglobins

The phenomena of gene crossing-over and gene duplication have been recognized for years in *Drosophila* on the basis of phenotypic observations made visually (Sturtevant and Beadle, 1962). It has recently become possible to observe these phenomena at the molecular level by studying the amino acid sequences of proteins that have resulted from

Table 6-13

Corticotropin	AcNH-ser tyr ser met glu his phe arg try gly lys pro val gly lys lys arg arg pro...
α MSH	AcNH-ser tyr ser met glu his phe arg try gly lys pro val-CONH$_2$
β MSH 1	NH$_2$-ala glu lys lys asp glu gly pro tyr arg met glu his phe arg try gly ser pro pro lys asp-COOH
2	NH$_2$-asp glu gly pro tyr arg met glu his phe arg try gly ser pro pro lys asp-COOH
3	NH$_2$-asp glu gly pro tyr lys met glu his phe arg try gly ser pro arg lys asp-COOH
4	NH$_2$-asp glu gly pro tyr lys met glu his phe arg try gly ser pro pro lys asp-COOH
Sheep lipotropic peptide	...glN ala ala glu lys lys asp ser gly pro tyr lys met glu his phe arg try gly ser pro arg lys asp lys arg tyr gly...

1 = human, 2 = monkey, 3 = horse, 4 = beef, pig.

such crossing-over and duplication, and in this manner some of the behavior of genes in evolution is revealed.

Such a study has been made by Smithies (1964) and others with respect to the haptoglobins. These are a group of proteins which are present in the blood serum of human beings and which combine with hemoglobin. The haptoglobins appear to have the antibodylike function of removing hemoglobin from the serum following its release from the red cells by a hemolytic incident. The combination of the two proteins is in the proportion of one molecule of hemoglobin with one molecule of type 1-1 haptoglobin. More than one molecule of hemoglobin may combine with the other two types, 2-1 and 2-2, of haptoglobin. The compounds formed by the combination are evidently eliminated from circulation by some physiological mechanism, and following a hemolytic crisis haptoglobin appears to be absent from the bloodstream.

The primary structure of the haptoglobins has been studied by Smithies and coworkers, and the results enable conclusions to be drawn regarding the evolution of these genetically variable molecules. The complete amino acid sequence of any one haptoglobin has not yet been determined, but enough is known of the various peptides that compose the molecules, including their partial amino acid sequence or content, to enable some important evolutionary conclusions to be drawn (*ibid.*).

It was found by Smithies, Connell, and Dixon (1961) that three types of haptoglobins were present in human beings; individuals could be designated as 1-1, 2-1, and 2-2 with respect to the haptoglobin type. Types 1-1 and 2-2 were homozygotes; 2-1 was heterozygotic with respect to containing haptoglobin 1 (short molecule) or 2 (long molecule). Further investigations showed that α and β chains were present in the haptoglobins and that there are three kinds of type 1-1 haptoglobin, termed $1^{\text{fast}}-1^{\text{fast}}$, $1^{\text{fast}}-1^{\text{slow}}$, and $1^{\text{slow}}-1^{\text{slow}}$, depending on the speed of their α chains in electrophoresis. The β chains of the haptoglobins are identical. The α chains, however, are variable, their differences being under genetic control.

The α chain of type 2 haptoglobin is about twice as long as that of type 1 and is composed of a type 1 fast (Hp1Fα) linked to a type 1 slow (Hp1Sα) chain with about 15 amino acids deleted at the point of junction. The relationships were summarized by Smithies as shown in Fig. 6-4.

In Fig. 6-4 N, F, J, S, and C designate chymotryptic peptides which are formed by enzymatic cleavage of the primary chain of a haptoglobin molecule at the points marked by arrows. Only the partial composition of the peptides is known, with the exception of F and S, and in this

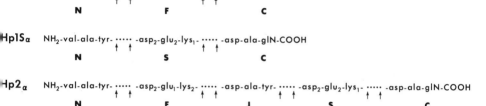

Fig. 6-4. The relationship of certain peptides in the haptoglobins (from Smithies, 1964).

case the sequence of the five amino acids is not yet known. J is stated by Smithies (1964) to be almost as long as one of the chains of insulin (20 to 30 amino acids). It is of much significance that J consists of part of C "grafted" on part of N.

Figure 6-4 shows first that the difference between Hp1Fα and Hp1Sα is a single amino acid interchange, lys/glu, in the pentapeptide designated as F or S (fast or slow). Hp1Fα and Hp1Sα are alleles, and the genetic difference between F and S can be attributed to the transitional change of a base pair in messenger RNA. The gene for Hp2α is formed by partial duplication or, more precisely, by end-to-end junction with partial overlapping of Hp1Fα and Hp1Sα genes in a heterozygote, so that peptide J in Hp2α has the composition partly of peptide C and partly of peptide N. The primary composition of Hp2α is that of a linear molecule consisting of Hp1Fα with about three amino acids lopped off the carboxyl end, joined to Hp1Fα with about twelve amino acids lopped off the amino end. It is as if the two words "tramping" and "trumping" were joined and condensed to form a new word "trampirumping" and as if this word has been adopted into the language! The switch point is at an alanine residue. The entire picture could scarcely be interpreted except in terms of collinearity between gene and protein. Furthermore, it is strongly indicative of an evolutionary process in which a larger and hence more complex gene is formed by splicing two simpler genes together, as shown in Fig. 6-4. We must also conclude that this is a comparatively recent event for the following reasons:

1. The amino acid composition of Hp2α corresponds to that of Hp1Fα *plus* Hp1Sα, allowance being made for the portion deleted by the overlap. Subsequent independent evolution of the Hp2α and the Hp1Fα (or its allele, Hp1Sα) genes should produce scattered differences in the respective amino acid sequences as in the analogous case of α and β hemoglobin chains. Such differences do not exist.

2. The Hp2α haptoglobin does not occur in primates other than human beings (Parker and Bearns, 1961), indicating that it is a recent arrival on the evolutionary scene.

Smithies has discussed the question of why there is not more than one Hp2α molecule, in other words, why the crossing-over of Hp1αF and Hp1αS always takes place at the alanine residue switch point mentioned above, when presumably a whole family of molecules could be produced by crossings-over taking place at various points. Among the possible explanations it would seem that the following could be considered:

1. The Hp2α molecule is not a useful protein unless it contains a J polypeptide of a certain length and primary structure; for example, this might be a necessity for forming the αβ haptoglobin combination.

2. There is some special property of the Hp1α genes that facilitates crossing-over at the alanine switch-point.

3. The gene Hp2α was formed only once as the result of a genetic accident; the explanation preferred by Smithies (1964).

In this regard, however, it must be recognized that other types of the Hp2 allele have been recognized by Nance and Smithies (1963) as follows: It is predictable that four types of Hp2α would result from the crossing-over of the Hp2α and Hp1α genes in heterozygous individuals. If we designate Hp2α as NFJSC and the two Hp1α alleles as NFC and NSC, the possible crossings-over would be

$$\frac{\text{NFJSC}}{\times \text{NFC}} \to \text{NFJFC (Hp2FF}\alpha\text{)} + \text{NSC (Hp1S}\alpha\text{)} \qquad (1)$$

$$\frac{\text{NFJSC}}{\times \text{NSC}} \to \text{NSJSC (Hp2SS}\alpha\text{)} + \text{NFC (Hp1F}\alpha\text{)} \qquad (2)$$

$$\frac{\text{NFJFC}}{\times \text{NSC}} \to \text{NSJFC (Hp2SF}\alpha\text{)} + \text{NFC (Hp1F}\alpha\text{)} \qquad (3)$$

$$\frac{\text{NSJSC}}{\times \text{NFC}} \to \text{NFJSC (wild type)} + \text{NSC (Hp1S}\alpha\text{)} \qquad (4)$$

The proteins Hp2FFα and Hp2SSα were discovered by Nance and Smithies in Brazil.

The next possibility involves the question of further crossing-over by the duplicated gene to result in triplication, and again Smithies, Connell, and Dixon (1962) have produced evidence for this in the haptoglobin system. Individuals homozygous for Hp2α ("2-2" individuals) should be subject to the following cross-over:

$$\frac{\text{NFJSC}}{\times \text{NFJSC}} \to \text{NFJFJSC} + \text{NSC (Hp1S}\alpha\text{)}$$

The protein NFJFJSC is present in the rare Johnson phenotype (Smithies, 1965). The primary structure of this protein has not been determined and is a question of great theoretical interest. It corresponds, in terms of the analogy previously used to a word written as "trampirampirumping."

The evolutionary significance of the Hp2α molecules was discussed by Smithies (1964). The gene controlling it is most common in India, where it predominates (Parker and Bearns, 1961). The enzyme heme α-methenyl oxygenase acts on the heme ring of hemoglobin during its degradation to biliverdin (Nakajima et al., 1963), and the most readily attacked substrate is hemoglobin combined with the homozygous

226 Other Proteins Showing Evolutionary Changes

Hp2-2. Hp1-1 and Hp2-1 are less effective in aiding the action of the enzyme. This phenomenon could presumably lead towards the predominance of a population of Hp2-2 individuals, but Smithies (1964) and Nance (1963) point out that such individuals will always be subject to genic crossing-over, which will produce some Hp1 genes. The unequal distribution of Hp2 genes throughout the world, however, indicates either that the equilibrium process has not reached completion, except possibly in India, or that the selective advantage of the 2-2 type varies in different parts of the world, depending perhaps on the relative frequency of hemolytic diseases.

A second example of crossing-over which has been studied at the molecular level is the case of the Lepore hemoglobins, which are hybrids between δ and β hemoglobins formed by unequal crossing over of the δ and β hemoglobin genes. These proteins are of the $\delta\beta$ type in that the NH_2 end of the molecule is followed by an amino acid sequence characteristic of δ hemoglobin. At points along the chain a switch takes place so that the molecule terminates with a sequence characteristic of β hemoglobin. The crossing-over point was located in hemoglobin Lepore$_{\text{Hollandia}}$ by Barnabas and Muller (1962). They found that two tryptic peptides corresponding to amino acid residues *9* to *17* and *18* to *32* had amino acid compositions identical with that of the δ chain of hemoglobin (Table 6-9). The next peptide, residues *33* to *42*, is the same in β and δ hemoglobins, but the next peptide after this, residues *43* to *61*, was identical in hemoglobin Lepore$_{\text{Hollandia}}$ and β hemoglobin. Therefore, the crossing-over had taken place in a locus in the gene corresponding to a point between residues *24* and *52*, since these are the locations of the β/δ differences in the peptides that mark the crossing-over. This may be represented as follows:

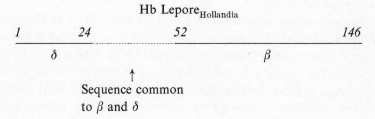

An interesting accompaniment of the hybridization of the molecule was that its rate of synthesis was greater than that of δ hemoglobin, which is normally produced at only 4 percent of the rate of β hemoglobin. The control of the two rates may thus be dependent upon the amino acid sequences in the respective chains of β and δ hemoglobin.

The Ferredoxins

An iron-containing protein, described as a link between hydrogenase and electron donors and acceptors, was isolated from *Clostridium pasteurianum* (Mortenson, Valentine, and Carnahan, 1962). A similar substance was found in spinach and the compound was stated to be the most electronegative electron carrier ($E_0' = -432$ mv at pH 7.55) that has been found in cellular oxidation-reduction reactions (Tagawa and Arnon, 1962). The compounds are termed *ferredoxins* (Mortenson et al., 1962).

Tagawa and Arnon found that *Cl. pasteurianum* ferredoxin ($E_0' = -417$ mv at pH 7.55) had spectral characteristics similar to those of the compound from spinach. Chloroplast ferredoxin had been known earlier in a partially characterized form as TPN-reducing factor, photosynthetic pyridine nucleotide reductase, haem-protein reducing factor, and cytochrome c photoreductase (Tagawa and Arnon, 1962).

Ferredoxin is present also in *Chromatium*, a photosynthetic microorganism that contains a hydrogenase, and has been shown to give a light-dependent evolution of hydrogen gas at the expense of a simple inorganic compound, such as thiosulfate. Ferredoxin is present in spinach chloroplasts as noted, but hydrogenase is absent. The addition of hydrogenase and ferredoxin from *Cl. pasteurianum* to chloroplasts enabled a photo-production of hydrogen gas to take place by linkage with the photochemical reaction system of the chloroplasts. The findings indicate a biochemical and evolutionary linkage between the photosynthetic reaction chain of the higher plants and that of anaerobic bacteria, the latter presumably retaining a vestigial remnant of the mechanisms for energy transformations carried out by primitive organisms at an early stage in the history of the earth when there was little or no oxygen in the atmosphere.

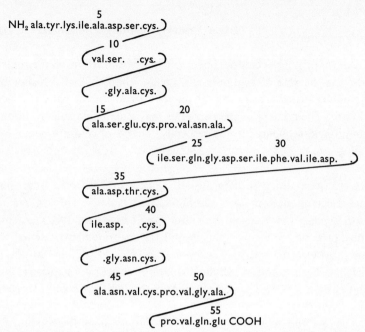

Fig. 6-5. Ferredoxin from *Cl. pasteurianum* (Tanaka et al., 1964) showing the incidents of partial gene duplication that appear to have taken place during evolution. The molecule appears to have developed from a linear pentamer of the tetrapeptide ala.gly.asn. cys (Jukes, 1966)

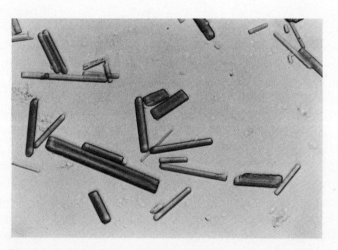

Fig. 6-6. Crystals of *Chromatium* ferredoxin (from Bachofen, Oda, and Arnon, personal communication).

A comparison of the amino acid sequences of the various ferredoxins should furnish a clue to an even earlier evolutionary divergence than is revealed by the cytochromes *c*. It will be of great interest to compare *Chromatium* (Fig. 6-5) and spinach ferredoxins.

The complete amino acid sequence of ferredoxin, the *Cl. pasteurianum* compound, was announced by Tanaka et al. (1964). The molecule is obviously an example of partial gene duplication. It may be represented as resulting from a nonadecapeptide which underwent four incidents of partial gene duplication, two deletions of a single amino acid, and the removal of a C-terminal heptapeptide by a point mutation which produced a chain-terminating triplet (Jukes, 1966). The nonadecapeptide in turn originated from five end-to-end condensations of a tetrapeptide, followed by loss of the C-terminal amino acid. The ferredoxin from *Cl. acidi-urici* contained arginine as a replacement for the lysine residue in ferredoxin from *Cl. pasteurianum*, and the *acidi-urici* compound was found to have one more proline and one more alanine than the *pasteurianum* counterpart, instead of phenylalanine and one of the serines in the latter (Buchanan, Lovenberg, and Rabinowitz, 1963). Crystals of *Chromatium* ferredoxin are shown in Fig. 6.6.

[7]

Homology and Deoxyribonucleic Acid

> Dich im Unendlichen zu finden
> Musst unterscheiden und dann verbinden.
> Goethe,
> Zeitschrift "Zur Naturwissenschaft
> überhaupt," 1822

Much attention has been given to the determination of the base composition of DNA in various organisms. Information on this subject was summarized by Belozersky and Spirin (1960) and by Marmur, Falkow, and Mandel (1963a).

The classification of sources of DNA on the basis of content of G + C is of very limited usefulness. It is somewhat analogous to classifying proteins on the basis of their content of glycine + alanine. Proteins that have the same content of glycine + alanine may be similar or different in all other respects; the only definite information from such a comparison would be that two proteins are different if they have different contents of glycine + alanine. As stated earlier, the G + C content will depend to a considerable extent on whether the organism uses coding triplets terminating in the majority of cases in G or C or prefers to use synonymous triplets terminating in A or U.

The next criterion for comparison would be nearest-neighbor analysis for the bases as described on page 30, Chapter 2. This criterion shows that the frequencies of the 16 possible dinucleotide pairs are unique and nonrandom in character for any given DNA sample. Certain bacteriophages show patterns similar to those of their hosts, but in most cases different organisms show different patterns of nearest-neighbor frequencies. The procedure is decidedly less sensitive for detecting homology than is the annealing procedure carried out with separated strands (Doty et al., 1960).

The final criterion of homology in the nucleic acids is a study of the base sequences in the molecules considered. A start has been made on this with the transfer RNA molecules, and the continuation of this study in various laboratories will undoubtedly supply much valuable information on taxonomic and evolutionary trends.

The hydrogen bonds that hold AT and GC together in DNA, and are hence responsible for "Watson-Crick" pairing between the two strands of DNA, may be parted by gentle heat, so that the two strands separate, or "melt." A considerable amount of strand breakage usually takes place during these manipulations. The extended rodlike character of the molecule is then lost, and the fragments of the two separate strands are free to form random coils. As noted elsewhere, it would seem highly unlikely that the long heterogeneous sequences of the individual strands could ever be rematched with the complementary pairs and could be once more united by hydrogen bonding that extended over any appreciable portion of the two halves of the molecule. This, however, has been shown to occur.

In 1960 it was reported that the two strands of DNA could be recombined in register by slow cooling after being separated by heating in solution. The recombination, a less ambiguous term for which is "annealing," was favored by increasing the ionic strength. It restored most of the A-T and G-C pairing and also the capacity to induce bacterial transformation (Doty et al., 1960; Marmur and Lane, 1960), which is regarded as an indication of double-strandedness. The annealing was by no means complete, as shown by the fact that the ultraviolet absorption characteristics of the recombined material differed appreciably from those of the native DNA (Marmur, Schildkraut, and Doty, 1961).

Further studies with isotopically labeled DNA showed that interchanging between the single strands of each pair of strands occurred freely during the process of separation and annealing. It therefore became possible to mix DNA preparations, AB and A'B', from two closely related organisms and to examine the percentage of cross-combination that would take place between strands A and B′ and strands A′ and B. This provided a means of examining the base-sequence homology between the DNAs of two closely related species.

Schildkraut and coworkers found a high degree of homology between DNA samples from the T-even bacteriophage series T2, T4, and T6 or from the T-odd bacteriophages T3 and T7. No cross-homology was found between T-even and T-odd bacteriophages (Schildkraut et al., 1962). This work laid the foundation for the extensive explorations of the DNA of various species by this procedure that have been made by McCarthy, Bolton, and Hoyer.

The DNA was prepared by Schildkraut and coworkers from bacteriophage T4 after labeling the growing culture with 5-bromodeoxyuridine (a heavy analog of thymidine) or with N^{15} + deuterium. The labeling procedures were used in density-banding to mark the formation of hybrid DNA. The usual procedure followed in detecting hybrid molecules was to prepare heavy DNA containing deuterium and N^{15} from one bacteriophage and mix it with unlabeled DNA from another bacteriophage. The mixture was heated to separate the strands and then cooled to recombine them, following which unrecombined single-stranded DNA was removed by hydrolysis with *E. coli* exonuclease I. The solution of the double-stranded molecules was examined by centrifugation in a cesium chloride gradient. This procedure led to the formation of three sharply defined peaks in the gradient when heavy and light T4 samples were mixed, melted, and cooled, the three peaks consisting, of course, of light-light, heavy-light, and heavy-heavy double strands. When heavy T4 and light T6 were submitted to the same procedure, the results were similar, and three peaks of light-light T6, heavy-heavy T4, and light-heavy T6-T4 were observed, but with T4 and T5 only two peaks (light-light and heavy-heavy) were produced, showing absence of annealing. Hybrid molecules were not formed between *E. coli* DNA and DNA from T3 and T7 bacteriophages, despite a suggestion that there may be a partial genetic homology between T3 DNA and *E. coli* DNA (Garen and Zinder, 1955). Bromouracil was effective for weighting the heavy strands and gave results similar to those with the heavy isotopes.

Evidence that DNA was a template for the biological synthesis of RNA accumulated following the discovery of DNA-like RNA by Volkin and Astrachan as discussed in Chapter 2, p. 34. At first the similarity of DNA and RNA was measured in terms of base composition, but this did not reveal whether complementary sequences were present. This question was examined by Hall and Spiegelman (1961), who combined equilibrium density centrifugation (Doty et al., 1960) with the use of two radioactive isotopes to demonstrate the formation of hybrids between single strands of T2-phage DNA and the corresponding DNA-like RNA. The latter would not form hybrids with the DNA of the host bacteria. Column chromatography was then examined as a means for detecting hybrid formation, and Bautz and Hall (1962) devised a procedure for preparing a DNA-cellulose compound by reacting glucosylated (T4-phage) DNA, dissolved in anhydrous pyridine, with acetylated phosphocellulose in the presence of dicyclohexyl carbodiimide. Bacteriophage-T4 messenger RNA could be isolated on T4-DNA-cellulose columns prepared in this manner and developed with saline citrate solution. The messenger RNA was then removed by raising the temperature and lowering the ionic strength and was found to behave as a template in an amino-acid-incorporating system.

The simpler procedure of entrapment on nitrocellulose membrane filters was then introduced by Nygaard and Hall (1963). These filters absorb single-stranded DNA of high molecular weight and RNA-DNA complexes but allow free RNA to pass through. Gillespie and Spiegelman (1965) added the step of drying the denatured DNA on the membrane filters for several hours before hybridizing with RNA. This prevented the re-formation of DNA-DNA complexes and enabled the single-stranded DNA to combine effectively with complementary RNA. Washing and treatment with pancreatic ribonuclease reduced the contamination with nonhybridized RNA to 0.003 percent of the input. The procedure was used in experiments which indicated that the "nucleolar organizer" segment of the X chromosome of *Drosophila melanogaster* contained a region that was complementary to ribosomal RNA (Ritossa and Spiegelman, 1965).

Discussions of DNA-RNA hybrids were presented in Chapters 2 and 3. The use of entrapment in agar for studying hybrids between two strands of denatured DNA and between DNA and RNA will now be described.

The hybridization of single strands of DNA from different organisms and the hybridization of DNA and RNA have been extensively used for taxonomic and other comparisons (Bolton and McCarthy, 1962; Hoyer, McCarthy, and Bolton, 1963, 1964; McCarthy and Bolton, 1963, 1964; McCarthy and Hoyer, 1964).

The general procedure is as follows:

1. A DNA sample of high molecular weight is heated to separate the strands by melting. The test solution is then cooled quickly to keep the strands from reannealing.

2. The separated single strands are mixed with a hot solution of agar and cooled rapidly. The gel is comminuted to granules. These granules contain trapped single-stranded DNA (sample 1), which is accessible to RNA or to other strands of DNA for complementation, because agar gel has a coarse molecular mesh that holds the strands of sample 1 in its cross-pieces while permitting access to them by the fragment of another sample. The granules are suspended in buffer and washed.

3. Another sample of DNA (sample 2) is labeled with P^{32} and broken into small fragments by shearing, and the fragments are then heated and cooled quickly to reduce them to the single-stranded state. These small fragments can diffuse into the agar granules and participate in pairing with complementary regions of the trapped DNA (sample 1). A representative condition for pairing is incubation at 60 C for 16 hours in 0.3 M NaCl. This was found by Hoyer et al. (1964) to favor a comparatively discriminatory type of trapping of sample 2.

Some of the strands of DNA sample 2 will recombine with each other in solution rather than annealing with the DNA in the agar, and the maximum amount that can be trapped in agar will reach a plateau dependent upon time, the proportion of DNA in agar to fragmented DNA, and other variables.

4. The granules are then washed and the hybridized radioactive strand is removed by warming or elution and the amount is measured by radioactive counting.

The amount of hybridization between two different DNAs is usually calibrated by referring it to the amount of homologous DNA that becomes bound. As an example, agar containing human DNA from HeLa cells bound 19.5 percent of a preparation of fragments of the

Table 7-1. Binding of Fragments of DNA, Human and Mouse Origin, by Single-Stranded DNA Strands Held in Agar

DNA in Agar		DNA Fragments Bound as Percent of Sample Labeled	
Origin	mg/g of Agar	Human, C^{14} 8γ	Mouse, P^{32} 15γ
Human	0.65	18	5
Mouse	1.02	6	22
Rhesus monkey	0.45	14	8
Rat	0.35	3	14
Hamster	0.37	3	12
Guinea pig	0.28	3	3
Rabbit	0.39	3	3
Beef	0.72	5	4
Salmon	0.60	1.5	1.5
E. coli	0.40	0.4	0.4
None	0	0.4	0.4

same DNA, H^3-labeled, as compared with 4.6 percent of a similarly prepared batch of mouse DNA, P^{32}-labeled; conversely, agar with mouse DNA bound 4.4 percent of human DNA and 18.3 percent of mouse DNA. A more extensive experiment of this type is shown in Table 7-1. A rough relationship is shown between DNA binding and the taxonomic relationship of the species that were compared.

An investigation was made of the type of homology occurring between species by studying the melting temperatures of various hybrid DNAs. The results indicated that dissociation of the hybrid pairs occurred at temperatures which could have been anticipated from the $G + C$ content. This indicated a high degree of pairing between bases and a low degree of looping out of unpaired regions in the double strand.

This finding tended to indicate that the annealing takes place between short, closely similar regions rather than representing a loose, scattered type of partial combination throughout the heterologous strands.

The interpretation of experiments of the type shown in Table 7-1 was that there exist certain base sequences in DNA that are shared by all vertebrates. These sequences represent about 20 percent of the DNA when various mammals are compared. Hoyer et al. (1964) pointed out that it was possible that the sequences represented regions that code for certain proteins not subject to much evolutionary change; examples would be the cytochromes c and pancreatic ribonuclease. Thousands of such proteins may exist but are unknown with respect to identity and composition.

It is also possible that such regions may include DNA base sequences that are not expressed phenotypically in adult species. The presence of a primitive morphology in the early mammalian embryo betokens the existence of genes for branchial arches, aortic arches, the mesonephros, and so on. These genes are presumably "turned off" as embryo development progresses. A "turning-off" process is specifically known to occur in the case of hemoglobin F, which is produced by the fetus during its sojourn in the uterus. Hemoglobin F has the composition $\alpha_2^A \gamma_2^F$. After birth takes place, it becomes gradually replaced by hemoglobin A, with the composition $\alpha_2^A \beta_2^A$, during the first six months of life. The γ polypeptide chains are no longer produced, because the genes that are responsible for their synthesis are apparently turned off, so that the messenger RNA that codes the γ chain is no longer transcribed, and its place is taken by another messenger that codes for the β chain.

The base sequences that code for the γ chain are still present in the DNA of adults, of course, and recombination of the strands would occur between preparations of DNA from an adult and a fetus if the samples were submitted to the Bolton-McCarthy procedure, another way of saying that the DNA of an individual remains constant with respect to composition and base sequence throughout life.

The annealing process between two single strands of DNA, each from one of two different but closely related organisms, may result from either of two causes. The first of these would be combination caused by a general similarity of the base sequence along the whole length of the strands. This would be analogous to a comparison of the similarity of the α hemoglobin chains of horse and human beings (Table 5-3), in which there are 17 amino acid differences scattered along two polypeptides, each containing 141 amino acids. The second possibility would be that the similarities and differences between the two DNA strands are bunched, regions of very close similarity being contrasted with regions of considerable differences. Apparently, the second possibility provides the best explanation of the observations, as shown in results described by Hoyer et al. (1965).

These investigators cross-reacted rhesus monkey and human DNA fragments. The monkey DNA was labeled with P^{32}, and the human DNA was labeled with C^{14}. The mixture was melted, cooled rapidly, and then incubated with agar containing human DNA in the presence of increasing amounts of unlabeled single-stranded fragments of monkey DNA. This procedure would tend to saturate the human DNA in the agar at all sites that would combine with monkey DNA. Nevertheless, a fraction of human DNA fragments continued to be bound by the DNA-agar without interference by the excess of monkey DNA. This leads to the comforting conclusion that there is a group of sequences in human DNA which are not represented by corresponding counterparts in monkey DNA. Let us hope that these sites represent the genes for our less simian characteristics, if such exist.

The same conclusion is emphasized by the results of other experiments. Cross-reactions were carried out between mouse DNA and human DNA. In the first experiments, agar containing human DNA bound 19.5 percent of human DNA fragments as compared with 4.6 percent of mouse fragments. In the second experiment, the conditions were reversed, so that agar containing mouse DNA was treated with the single-stranded fragments; in this case 4.4 percent of human DNA and 18.3 percent of mouse DNA were bound to the agar. Hoyer and coworkers concluded from this experiment that the two species appeared to have between 20 and 25 percent of their DNA sequences in common.

The taxonomic relationship between groups of animals was compared with the competitive reaction in the binding of DNA fragments. The procedure is as follows: Agar was impregnated with rhesus monkey DNA. A small amount (0.5 mg) of P^{32}-labeled rhesus monkey DNA fragments was incubated with a large excess of the DNA agar in the presence of varying quantities of unlabeled DNA fragments from various sources. As was anticipated, a large excess of rhesus monkey unlabeled DNA inhibited the binding of the labeled rhesus monkey DNA. The binding was equally inhibited by baboon DNA, strongly inhibited by the DNA of human beings, chimpanzees, or owl monkeys, moderately inhibited by that of Galago or Potto monkeys, slightly inhibited by hedgehog or mouse DNA, and very slightly by chicken DNA. The results indicate a taxonomic separation as reflected by an increasing diversity of base-sequence grouping in DNA, this representing, in turn, increasing differences in the proteins of species involved. Such differences, of course, find their counterpart in immunology.

The DNA-agar procedure has been used in studies of complementary binding of DNA and RNA, and the results were comparable to those obtained by other methods. The general view of the relationship between DNA and RNA is that all RNA (except viral RNA) is produced in vivo by the RNA polymerase reaction, in which one of the two DNA strands is used as a template and three types of RNA—ribosomal, transfer, and messenger—are formed by this reaction. Since the ribosomal and transfer RNA molecules are extremely short in comparison with the overall length of the DNA molecules in any organism, most of the DNA constitutes the information for the transcription of messenger RNA, which is produced in varying quantities in the cell because of control mechanisms. Most of the RNA in a cell is ribosomal RNA, which will complement only a small portion of the DNA base sequence, so that this portion may be quickly saturated in an agar column experiment with a negligible fraction of the ribosomal RNA.

McCarthy and Bolton (1964) found that their agar columns containing single-stranded DNA could bind RNA from the same species of organism, *E. coli*. Two preparations of RNA were used, the first from cells grown for four generations on C^{14} uracil and the second from cells exposed to P^{32} for 2 minutes before harvesting. The mixture of these two preparations was allowed to complement with the DNA-agar under conditions optimal for annealing and the bound RNA was then eluted. A fraction was obtained containing 1.2 percent of the C^{14} and 21 percent of the P^{32}. This fraction should represent messenger RNA if one accepts the concept that the DNA binding site for ribosomal RNA is negligible in size (estimated as 0.2 percent of the total) compared with the messenger-binding sites and that the percentage of messenger RNA in total RNA is therefore about 1.2 percent, it being assumed that transfer RNA (double-stranded) did not bind under the conditions of the experiment. The procedure can be used for purifying messenger RNA.

The question of whether only one of the two strands of DNA is involved in RNA complementation was easily examined by the agar column technique as follows: If only one strand of DNA is complementary to RNA, then only half of the DNA on an agar column would anneal with an excess of RNA, and the remaining half, representing the strand homologous to the RNA, would remain open for combining with DNA fragments, as shown in Fig. 7-1. This was found

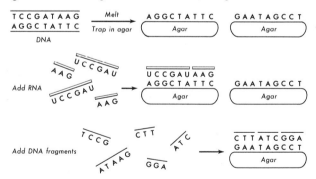

Fig. 7-1. Diagram showing how the agar-column procedure is used to demonstrate that only one strand of DNA binds RNA.

to be the case. A DNA-agar preparation was saturated after treatment with 13 mg of RNA. It was then incubated with fragmented DNA, and 14.3 µg of DNA was bound. An identical preparation was incubated with DNA without RNA pretreatment, and 28.5 µg of DNA was bound. Half the possible sites were therefore occupied with RNA in the first experiment. The DNA and RNA preparations were total fractions obtained from *E. coli* BB. Labeling was with P^{32}.

Classification of Bacteria as Related to the Base Composition of DNA

There are several methods for determining the base composition of a sample of DNA in terms of the percentage content of each of the four bases. Chemical analyses may be used, but it is more convenient to employ one of the various physicochemical methods, such as the thermal denaturation temperature (T_m) or the cesium chloride density gradient technique, which is dependent upon the equilibrium position taken up by a sample of DNA following prolonged ultracentrifugation in a gradient of cesium chloride solution. The results are commonly expressed in terms of percent G + C in the total bases. Since G = C and A = T, the base composition can be calculated from the value of G + C.

It is obvious that two species with an equal percentage of G + C in their DNA may have similar or quite different base sequences, and this may be closely related or not related at all. On the other hand, two organisms with DNA containing pronouncedly different percentages of G + C will certainly differ in other respects, although, as noted previously (Chapter 2), differences in G + C may be far greater than differences in amino acid composition of the total protein coded because of synonymities among the coding triplets. Each amino acid except methionine and possibly tryptophan has at least one coding triplet ending in G or C and at least one other ending in A or U.

The base composition of the DNA of numerous species of bacteria has been examined by various investigators, including among others, Belozersky and Spirin (1960), Sueoka (1961a), Marmur and Lane (1960), and Marmur, Falkow, and Mandel (1963a). In their review,

Marmur et al. (1963a) compiled data on base composition for more than 200 bacteria and arranged them in ascending order of G + C content at intervals of percent G + C. A continuous spectrum is shown by this tabulation, starting with low G + C content in the *Clostridia* and ranging upwards through *Streptococci* and *Hemophili*, through *Salmonellae* and *Corynebacteriae*, and to *Pseudomonads* and

Table 7-2 Percentage of G + C in DNA of a Representative Series of Organisms That Show Differences between Species

Organism	mol % G + C in DNA
Clostridium perfringens, *Clostridium tetani*	30–32
Staphylococcus aureus	32–34
Bacillus cereus, *Treponema pallidum*	34–36
Diplococcus pneumoniae, *Hemophilus influenzae*	38–40
Higher animals and plants	38 ± 10
Neisseria catarrhalis, *Leptospira biflexa*	40–42
Bacillus subtilis, *Bacillus stearothermophilus*	42–44
Saprospira grandis, *Desulfovibrio orientis*	44–46
Vibrio comma, *Pasteurella pestis*	46–48
Spirillum serpens, *Neisseria gonorrhoeae*	48–50
Escherichia coli, *Salmonella typhimurium*	50–52
Aerobacter cloacae, *Corynebacterium diphtheriae*	52–54
Aerobacter aerogenes, *Brucella abortus*	54–56
Azotobacter vinelandii, *Lactobacillus bifidus*	56–58
Agrobacterium tumefaciens, *Serratia marcescens*	58–60
Pseudomonas fluorescens, *Rhodospirillum rubrum*	60–62
Xanthomonas pelargoni, *Arthrobacter globiformis*	62–64
Flavobacterium vitarumens, *Desulfovibrio desulfuricans*	64–66
Halobacterium salinarum, *Mycobacterium tuberculosis*	66–68
Sarcina flava, *Pseudomonas saccharophila*	68–70
Micrococcus lysodeikticus, *Streptomyces griseus*	70–80

Source: Marmur et al. (1963a).

Streptomyces, to name but a few genera. An abbreviated list is in Table 7-2. The findings may imply the existence of some kind of an evolutionary gradient.

The transformability of *B. subtilis* by the DNA of other organisms was confined to members of the genus *Bacillus* with DNA base composition similar to that of *B. subtilis*. Differences in base sequences may exist, however, even when two DNAs have similar G + C content and when, in addition, the two organisms can conjugate. This takes place by cell contact and transfer of most or all of the DNA from the donor to the recipient. A case in point is the transfer of genes, including those for lactose utilization and indole production, from *E. coli* K12 to *Salmonella typhimurium* (Brinton and Baron, 1960). Both organisms have DNA containing 51 ± 1 percent $G + C$, and the order and spacing of genes are identical (Falkow, Rownd, and Baron, 1962). The DNAs will not hybridize when subjected to denaturing and slow cooling in the usual procedure for obtaining recombination (*ibid.*) This implies that the annealing test is more specific for detecting homology than is the conjugation test. Similarly, transformation and transduction in bacteria appear to be more specific than conjugation and to be well correlated with DNA hybrid formation (Schildkraut et al., 1962; Marmur, Seaman, and Levine, 1963*b*).

The proteins of a number of species of microorganisms were analyzed for their amino acid content by Sueoka (1961*b*), who compared the percentages of various amino acids with the base composition of the DNA. Certain correlations were noted; $A + T$ trended upward with lys and ilu and downward with ala and arg. The coding assignments subsequently made with synthetic polyribonucleotides were in accordance with these observations; lys and ilu have codes predominantly high in A and U, while ala and arg are coded by triplets containing C and G.

Homology and Nearest-Neighbor Frequencies

The frequency of base pairs in the DNA of various organisms has been determined with great care by using the nearest-neighbor procedure, by Josse, Kaiser and Kornberg (1961) and Swartz, Trautner, and Kornberg (1962). The results fully confirmed the model of the DNA molecule based on Watson-Crick pairing and antiparallel strands, as noted in Chapter 2. Can any further conclusions be gleaned from the nearest-neighbor data?

Let us consider the base sequence of the genome of an organism in terms of the simplest analogy, that of the continuous succession of words in a book written in English. The analogy is inadequate in many respects: the alphabet has 26 letters instead of the 4 letters in DNA, vowels predominate—but this can be corrected by a calculation in which the number of pairs of two letters such as *a* and *b* is divided by the total number of *a*'s multiplied by the total number of *b*'s—the words in a book are much shorter than the sequences of about 800 bases that spell out the molecule of a protein monomer, and so on. The analogy is useful, however, for drawing some comparisons. In both cases we can look for nearest-neighbor sequences. These will be unrestricted with respect to the 16 possible pairs in DNA, but in English *q* will always be followed by *u*. The distribution of nearest-neighbor pairs in DNA has been shown to be nonrandom and to be characteristic for a species. We would expect this also to be the case with the pairs of letters in a book. The repeated use of certain words would tend to make some pairs of letters more abundant than others. The exact totals of each pair in the book would almost certainly make it different from any other book and would be useful in characterizing the book if we had no other way of so doing. Two textbooks of organic chemistry, written by different authors, would probably show more similarity to each other by this test than would either one to a novel.

The nearest-neighbor data, or numbers of each of the 16 base pairs, furnish a rough test for similarity of samples of DNA but no information on their protein coding properties. A base pair may be within a coding triplet, or it may bridge two triplets. Furthermore, coding functions have been assigned to all the 64 triplets (Table 2-12). The nearest-neighbor percentages will undoubtedly be related to the amino acid composition of the proteins of our organism but in a manner that defies prediction in the foreseeable future.

To summarize, the experimental formation of hydrogen-bonded hybrid double strands by fragments of single-stranded DNA, produced by denaturation, is correlated with genetic homology and corresponds roughly to taxonomic relationships. The experimental tests with DNA-DNA hybridization were initiated by studies in density gradients and since then have been carried out for the most part with DNA entrapped in agar granules. DNA-RNA hybrids have been studied by these procedures and by the use of cellulose columns or by retention on nitrocellulose membrane filters, which permit the passage of RNA. The findings have shown that RNA is complementary to only one strand of DNA and that regions of complementarity for messenger, ribosomal, and transfer RNA are present in homologous DNA. Some heterologous bonding of RNA is shown by DNA from closely related species.

[8]

Biochemistry and Evolutionary Pathways

> Desinet ante dies et in alto Phoebus anhelos
> aequore tinguet equos, quam consequar omnia verbis
> in species translata novas.
> <div align="right">Ovid, <i>Metamorphoses</i> xv, i. 418</div>
> Evening has overtaken me, and the sun has dipped below
> the horizon of the Pacific Ocean, yet I have not had
> time to tell you of all the things that have evolved
> into new forms.

It is not the purpose of this book to stray from the molecular pathway into the vast jungles of complexity that are to be found in protoplasm. It may well be that the only flawless way to characterize a species is to write the entire base sequence of its DNA and simultaneously to read off the amino acid sequences of the thousands of proteins that are coded by it. Going a step further, these sequences govern the secondary, tertiary, and quaternary structures of the proteins, and presumably the properties of enzymes could also be predicted from their amino acid sequences. Perhaps some day these matters will be fully computerized. For the time being, classification is best left to the taxonomists. We do not propose to state that a cow and a pig are members of the same species because their cytochromes c are identical; this is akin to declaring that a man with white hair and a beard is a mountain goat (*Oreamnos montanus*). There are, nevertheless, some interesting matters relating to evolution that will be reviewed in this chapter even though they are concerned with proteins whose chemical structure is unknown.

The simplified train of events commonly proposed for biological evolution consists of three steps: first, the appearance of life in the "primeval nutritional soup" which supplied all the necessary food ingredients; second, depletion of the food supply and development of biologically synthetic functions, including photosynthesis; third, loss of certain of these functions by parasitic, saprozoic, and saprophytic organisms, together with the continued development of photosynthesis by the green plants.

Cohen (1963) pointed out that the possibility of a relatively recent gain of structure and function should be added to this simple evolutionary pattern. He has remarked that the example of the antibiotics represents "an astonishing new world of apparent biochemical freaks." It is significant that more than one pathway of biological synthesis may exist for an important metabolite such as lysine or arginine. Such duplication is not embraced in the evolutionary pathways arising in the manner postulated by Horowitz (1945), so that his theory should be supplemented by provision for biochemical innovations that give rise to alternate metabolic pathways. This is quite consistent with gene duplication followed by differentiation.

Alternate pathways may occur in the same organism and have an advantageous effect; for example, methyl groups of choline can originate in rats and other species by transmethylation from S-adenosylmethionine, which is then regenerated from S-adenosylhomocysteine by the action of folic acid and vitamin B_{12}.

When the diet is deficient in vitamin B_{12} and folic acid, choline can receive its methyl groups from betaine by transmethylation. This alternate pathway may be shown by suitable experiments to protect young rats from dying with the hemorrhagic kidney syndrome that is due to choline deficiency (Griffith and Mulford, (1941). Thus, an alternate pathway may confer benefits that could aid in survival of a species.

Two biochemical pathways exist for the synthesis of lysine. The two do not occur simultaneously in any single organism. Vogel pointed out that the manner in which organisms obtain their supply of lysine is useful in the examination of evolutionary relationships because it is "suitably poised between the extremes of biochemical unity and diversity" (Vogel, 1965; van Niel, 1949).

Parasites and saprophytes, including all the multicellular animals, obtain lysine as a preformed molecule from their food, and the need for this amino acid probably forms part of the background of the carnivorous habit of predators. The higher plants, together with the blue-green and green algae, synthesize lysine through a well-defined and distinctive metabolic pathway involving at least seven enzymatic steps that start with the condensation of L-aspartic β-semialdehyde and pyruvic acid and end with the decarboxylation of diaminopimelic acid (DAP). This may be termed the *DAP pathway* (Fig. 8-1*a*).

Up to this point the story is typical of many food substances that are synthesized by plants and eaten by animals. In the case of lysine, however, there is a second metabolic pathway which is quite different from the DAP pathway. This is the α-aminoadipic (AAA) acid pathway of biosynthesis for lysine, starting with α-ketoglutaric acid and acetyl coenzyme A and followed by eight enzymatic steps (Fig. 8-1*b*). The AAA pathway is used by many of the higher fungi and by Euglenids. The distribution of the DAP and AAA routes shows a sharp taxonomic demarcation.

By the ingenious use of certain specifically labeled radioactive precursors, Vogel found it possible to distinguish sharply between the two lysine pathways. The precursors used were 3- and 4-C^{14}-labeled aspartic acids, 1-C^{14}-labeled alanine and 2-C^{14}-labeled acetate. The organism was grown on a culture medium containing these tracers, and the radioactivity of aspartic acid, threonine, and lysine in the culture was then measured. These three metabolites had approximately the same radioactivity when an organism using the DAP pathway received 3- or 4-labeled aspartic acid or 3-labeled acetate. With 1-labeled alanine the radioactivity of lysine was higher than of aspartic acid or threonine. In contrast, organisms using the AAA pathway did not incorporate radioactivity into lysine from 4-labeled aspartic acid or 1-labeled alanine but did incorporate radioactivity into lysine from 3-labeled aspartic acid and especially from 2-labeled acetate. These differences arise from the disparity between the two synthetic pathways, and the biochemical interpretations of the intermediate reactions were explained by Vogel. He reported that all bacteria tested used the DAP

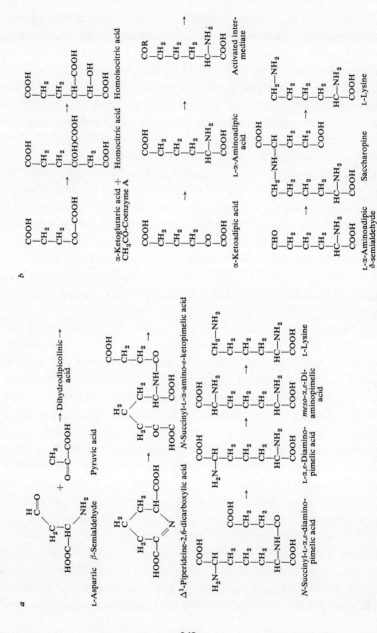

Fig. 8-1. (*a*) Diaminopimelic acid pathway. (*b*) α-Aminoadipic acid pathway of lysine biosynthesis.

lysine pathway. This was also the case for two blue-green algae and two green algae and for a fern, a gymnosperm, and angiosperms, which included two monocotyledons and two dicotyledons.

The AAA pathway occurred in all the ascomycetes and basidiomycetes that were examined. There was a division among the phycomycetes. Those producing posteriorly uniflagellate or nonflagellate

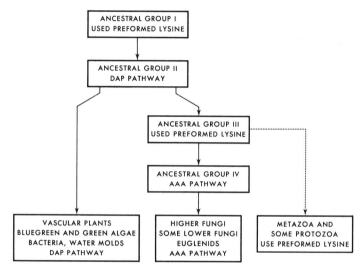

Fig. 8-2. Postulated evolutionary relationships in lysine pathways (from Vogel, 1965).

spores use the AAA pathway, while those whose spores were anteriorly uniflagellate or biflagellate synthesized lysine via DAP. This division of the phycomycetes by a biochemical characterization is indeed remarkable.

The observation that yeast uses the AAA pathway, thus placing it closer than the green plants in its relation to animals (Fig. 8-2) is of much interest, for yeast cytochrome c has marked resemblances to mammalian cytochrome c. In Fig. 8-2, the green plants are further removed than yeast from the mammals. This would lead one to anticipate that wheat germ cytochrome c would differ more widely than yeast cytochrome c from mammalian cytochromes c.

250 Biochemistry and Evolutionary Pathways

Vogel considers that the two pathways were developed separately in organisms originally incapable of lysine synthesis and that the DAP pathway is more ancient because it occurs in many relatively primitive bacteria, including pseudomonads, *Eubacteriales*, and actinomycetes. This would indicate that the DAP pathway was developed by archetypal organisms following the exhaustion of the supply of lysine from the "primeval nutritive soup" (p. 64). An offshoot then took place in the emergence of an ancestral group of parasites and saprophytes which lost their ability to synthesize lysine and instead subsisted on the lysine that was made by their more industrious neighbors. According to Vogel's scheme, some of this group of parasites and saprophytes then developed a new pathway which made them independent of preformed lysine. This was the AAA pathway, which Vogel regards as a "biochemical innovation," as defined by Cohen (1963).

We might include the existence of a fourth pathway that has been developed by *Homo sapiens* within recent years, namely, the synthesis of lysine by the techniques of organic chemistry (Eck and Marvel, 1934; Rogers, 1951, 1952). This is not a biological pathway in the conventional evolutionary sense, but it should be noted that the technological achievements of the human race in chemically synthesizing organic nutrients such as thiamine, riboflavin, nicotinic acid, and lysine is having a profound effect upon increasing the numbers of people and thus upon the terrestrial environment.

Evolutionary Pathways and Vitamin A.
Metamorphosis

The carotenoids are a group of plant pigments which include α, β, and γ carotene, cryptoxanthin, lycopene, and xanthophyll. The formula for β-carotene is as follows:

$$\text{β-carotene structural formula}$$

Hydrolysis of this molecule at the midpoint, with the addition of two molecules of water, gives two molecules of vitamin A (Fig. 8-3), but α and γ carotene and cryptoxanthin contain only one rather than two rings characteristic of the β-ionone group,

$$\underset{\beta\text{-ionone}}{\underset{}{\text{[ring with CH}_3\text{, CH}_3\text{, CH}_3\text{ substituents]}-CH=CH-\overset{O}{\underset{}{C}}-CH_3}}$$

and so can give rise to only one molecule of vitamin A_1. Lycopene and xanthophyll do not contain the ring found in β-ionone and do not give rise to vitamin A_1. The biological synthesis of the carotenoids is carried out by green plants, but plants do not make vitamin A_1. This is produced enzymatically from the first four carotenoids mentioned above by the organisms in certain phyla in the animal kingdom, including the crustacea, the arthropods, and the vertebrates. Animals cannot synthesize the carotenoid precursors of vitamin A, but they can utilize preformed vitamin A obtained by predation.

Wald, whose interest in evolution and the origin of life (1964) is well complemented by his experimental discoveries in the biochemistry of vision (1961), has combined these fields in a study of evolutionary pathways in which vitamin A is involved. In so doing, he has provided illustrations, which we shall discuss, of the relationships between molecular processes and biological events (1963).

Vitamin A_1 is used for *vision* by all animals which contain it, and it has additional functions in the vertebrates. The visual cycle includes the dehydrogenation of vitamin A_1 to retinene, the combination of retinene with the protein *opsin* to form rhodopsin, the bleaching of rhodopsin by light, and its regeneration in the dark (Wald, 1963).

The enzymatic reaction by which vitamin A_1 is produced from the carotenoids in animals is unknown, but its appearance was evidently a key step in evolution, since the vitamin is essential in all image-resolving eyes. It is found in two forms, vitamins A_1 and A_2 (Fig. 8-3); A_2 is produced from A_1 by an enzymatic reaction whose phylogenetic distribution will be discussed later. Vitamin A_2 forms porphyropsin, a

252 Biochemistry and Evolutionary Pathways

visual pigment corresponding to rhodopsin. Porphyropsin and rhodopsin contain retinene in the 11-*cis* configuration, which possesses specific photochemical properties associated with its function in the eye.

The conversion of vitamin A_1 to A_2 occurs in certain vertebrates: it takes place by the removal of the two hydrogen atoms (Fig. 8-3). The two forms are distributed among vertebrates in terms of their habitat. The freshwater vertebrates use vitamin A_2 in vision; the marine fishes, the birds, and the mammals use vitamin A_1. The distribution of the vitamins in various animals has been studied by Wald (1963) who noted that the relation of A_1 and A_2 to habitat extends to the reptiles; the alligator, the rattlesnake, and marine turtles have A_1, while freshwater turtles have A_2. An interesting question arises with respect to the euryhaline fishes, which live as adults both in the sea and in fresh water. These fishes contain a mixture of A_1 and A_2. Wald has shown that the proportion of A_1 to A_2 among these is related to whether they spawn in fresh or salt water; A_2 predominates in fishes such as salmon and trout, which spawn in fresh water, but the freshwater eels, which return to the sea to spawn, contain more vitamin A_1.

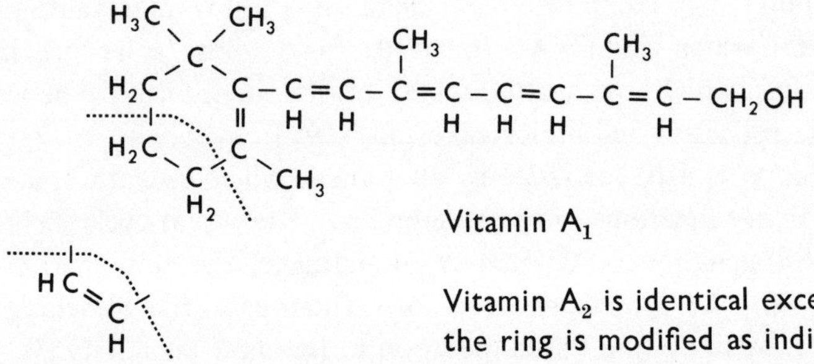

Vitamin A_1

Vitamin A_2 is identical except that the ring is modified as indicated.

Fig. 8-3

Wald's proposal is that the primitive vitamin A of vertebrates is A_2 as shown by its association with freshwater vertebrates. The mechanism for converting A_1 to A_2 seems to have been lost or "shut off" by the "newer" vertebrates, which migrated during evolution from fresh water to terrestrial and marine habitats. Furthermore, some of the amphibians, as exemplified by the bullfrog, change from A_2 to A_1 at metamorphosis, when they change from aquatic to terrestrial form.

It is familiar to biologists that the evolution of living organisms has left its imprint on the steps that take place during their embryonic development. This phenomenon was used to great advantage by Darwin (1859) in drawing attention to the recapitulation of ancient and extinct forms that occurs during the embryonic or larval period. Various "primitive" stages are exhibited during the embryonic growth and differentiation of the organs and tissues of young multicellular animals. As summarized by Storer and Usinger (1965),

A fish embryo develops paired gill slits, gills, aortic arches, and a two-chambered heart; these all persist in the adult to serve in aquatic respiration. Comparable structures appear in a frog embryo and are necessary during the fish-like life of the frog larva in water. When the larva transforms into an air-breathing frog, however, the gills and gill slits disappear, lungs become functional for respiration in air, the aortic arches are altered to serve the adult structure, and the heart is three chambered for circulation of the blood to both the body and lungs.... Astonishingly, the early embryos of reptiles, birds, and mammals also develop a fish-like pattern of gill slits, aortic arches, and two-chambered heart, although none of them has an aquatic larva and all respire only by lungs after birth. The embryonic gill slits soon close; the multiple aortic arches become the carotids and other arteries; and the heart soon becomes three-chambered, later having four chambers in birds and mammals.

The presence of gill slits and multiple aortic arches in embryos of reptiles, birds, and mammals is not explained by a theory of special creation, but under a theory of evolution they are obviously ancestral relics. The fossil record indicates that aquatic, gill-breathing vertebrates preceded the air-breathing land forms. In point of time their sequence of appearance was fishes, amphibians, reptiles, birds, and mammals.

At a later period in life, sometimes suddenly, the changes of metamorphosis take place. Embryonic development and metamorphosis are fruitful subjects for exploration in terms of the molecular changes that engender them, and many discoveries of evolutionary significance undoubtedly lie ahead in this field for biochemists. Today we assume that the "ancestral relics" in embryonic life are coded by base sequences in DNA that are not transcribed by adults.

Wald (1963) has described "second metamorphosis" as reversing in part the changes of the first metamorphosis and as completing the life cycle by bringing the organism back to the beginnings of the next generation. In the whole organism, this is illustrated by the return of the salmon to its spawning grounds; in cellular terms, the mammalian spermatozoon and ovum travel to the uterine endometrium to complete the cyclical journey of life and to initiate the formation of a new member of the species.

The change from vitamin A_2 to A_1 in the visual cycle at metamorphosis in the bullfrog is quite complete and, like the anatomical changes of metamorphosis, can be engendered prematurely by treatment with thyroxine (Wilt, 1959). The disappearance of A_2 is apparently due to the cessation of the conversion of A_1 to A_2 in the retina (Wald, 1963). Simultaneously, there is a change from fetal to adult hemoglobin; this biochemical event occurs in all classes of vertebrates that have been investigated, and Moss and Ingram (1965) found that it may also be produced in frog tadpoles by treatment with thyroxine. The excretion of ammonia, characteristic of fishes, becomes changed to the excretion of urea that is seen in terrestrial animals. Wald (1963) points out that the anatomical and biochemical changes that occur during the embryology of land vertebrates may be regarded as vestiges of the metamorphosis that exists in amphibians. The terrestrial animals live out their prenatal days in the aquatic surroundings that resemble those of the tadpoles of today or of our ancestors of a vanished era. The changes of metamorphosis, like those of evolution, are biochemical. In contrast to evolution, however, metamorphosis consists of the release of biochemical processes by hormonal action; the dormant genetic instructions are activated by the formation of RNA. These instructions represent the long accumulation of evolutionary additions to the DNA while the primitive or larval information was retained for embryonic use. Examples of the phenomenon of "second metamorphosis" are seen in conspicuous forms in newts (*Triturus viridescens*), which spend their larval form in the water, then emerge to land, and later return to an aquatic life to spawn and spend the rest of their lives. The onset of the second metamorphosis is inducible by prolactin, the lactogenic hormone of the anterior pituitary. The effect is presumably to suppress the genes that code for the biochemical changes of the first metamorphosis and to bring about the expression of certain genes that are associated with the mechanisms of the larval stage. These include the excretion of ammonia, and the production of vitamin A_2 (Wald, 1963). Wald has provided further illustrations of the biochemical changes accompanying the second metamorphosis. In freshwater eels, vitamin A_2 disappears from the mixture of A_1 and A_2 as the animals prepare to migrate to the ocean for spawning, while the reverse process takes place in the sea lamprey, which contains vitamin A_1 during its downstream migration to salt water and changes to vitamin A_2 when it returns upstream to spawn.

The approximate generalization may be made that many biochemical processes that date from an early period in evolution have been preserved in the genome for use during embryonic life. The genes that code for the phenotypic expression of these processes are typically activated for transcription by prolactin. In contrast, there is a group of biochemical processes that are of recent evolutionary origin and are related to adult life. These are coded by genes whose expression is stimulated by thyroxine; simultaneously, it appears that thyroxine may repress the genes that are responsible for the "earlier" processes and that respond to prolactin.

Taxonomically Related Difference in Enzymes and the Products of Their Reactions

Immunological reactions, including the precipitin test and the complement fixation reaction, have long been used in the exploration of taxonomic relations among different species (Nuttall, 1904). In recent years, these techniques have been expanded so that immunoelectrophoresis is now used, as in the studies by Manski, Auerbach, and Halbert (1960; also Manski, Halbert, and Auerbach, 1961), who have related the results of precipitin tests and the immunoelectrophoretic patterns obtained with lens proteins to the taxonomy of various vertebrates. Such procedures can be assumed to indicate similarities and differences in the amino acid sequences among a group of homologous proteins obtained from several species. The soluble proteins of the lens include 10 or more antigenic components. Antilens sera were prepared by injecting a soluble lens protein into a rabbit, and the rabbit antiserum was then tested by two-directional gel diffusion and immunoelectrophoresis against lens proteins. Such experiments show cross-reactions between an antibody and a related protein in the form of a ring of precipitated protein where the two components react in an agar gel.

Numerous other investigations of this type have been carried out with various groups of animals by the use of similar immunological procedures and also by the study of the electrophoretic patterns of mixtures of proteins obtained from various tissues. Many of these studies were reported in a recent symposium volume (Leone, ed., 1964). A discussion of them is beyond the scope of this book.

Table 8-1. Rates of Reaction of the 3-Acetylpyridine Analog of DPN with Different Rabbit Dehydrogenases

Dehydrogenase	Rate of AcPyDPN Reaction Compared with DPN (percent)
Liver alcohol	450
Liver glutamic	150
Heart mitochondrial malic	125
Muscle lactic	22
Muscle triosephosphate	10
Heart lactic	4
Liver β-hydroxybutyrate	<1
Muscle α-glycerophosphate	0

Kaplan and his collaborators (1965) have studied the catalytic properties of enzymes obtained from different species, the dehydrogenases that are concerned with the oxidation of nicotinamide adenine dinucleotide ($NADH_2$) in the transfer of hydrogen from this coenzyme to various metabolites. Hundreds of such enzymes probably exist in each of many living organisms, thus indicating that many of these various enzymes might each possess similar binding sites for the accommodation of the coenzyme. Such families of enzymes might result from repeated gene duplication during evolution, or each might evolve independently if the binding site for $NADH_2$ is simple and flexible with regard to number and varieties of amino acids involved.

One approach by Kaplan's group was to compare a family of

dehydrogenases, found in rabbits, with respect to their rates of reaction with an analog of $NADH_2$, 3-acetylpyridine. The comparative results that were obtained are in Table 8-1. The results indicate differences in the binding sites of the respective enzymes. This might well indicate that the coenzyme itself can react with a number of different active sites. Extensive studies of amino acid sequences are needed to interpret this finding. Fingerprinting methods with trypsin hydrolysates are of very little value for this purpose, because while they reveal changes in lysine and arginine residues, they tell little about other sites.

Table 8-2. Amino Acid Compositions of Heart- and Muscle-Type Lactic Dehydrogenases from Chicken and Beef

Amino Acid	Chicken		Beef	
	H_4	M_4	H_4	M_4
ala	88	81	80	78
arg	35	35	34	42
asN, asp	129	125	132	127
cys	26	26	17	26
glu, glN	122	102	129	121
gly	96	104	98	100
his	30	63	26	33
ilu	66	85	85	91
leu	149	121	143	136
lys	99	112	96	103
met	25	31	36	32
phe	19	27	21	29
pro	38	44	46	51
ser	107	110	97	87
thr	75	51	56	48
try	22	24	22	24
tyr	31	19	29	29
val	125	121	138	115

Source: Kaplan (1965).

Kaplan found that there were two main types of $NADH_2$-linked specific lactic dehydrogenases; one type (H) was found in heart muscle and the other (M) in skeletal muscle. Both types are tetrameric molecules with a molecular weight of about 140,000. The approximate composition of each type obtained from two different sources is shown in Table 8-2, in which H_4 and M_4 denote the tetramers (Kaplan, 1965). Similarities of the amino acid sequences of H and M from one species lead one to suspect that they might have originated by gene duplication followed by subsequent evolutionary modifications. Furthermore, the amino acid composition of all four of the proteins shown in Table 8-2 shows a similarity which would be compatible with a derivation from a common archetype by the combined processes of gene duplication and separation of species. It is not possible to examine this point adequately

Table 8-3. Amino Acid Composition of M_4 Lactic Dehydrogenase

Species	Histidine	Phenylalanine
	(residues/mole)	
Chicken	63	27
Pheasant	61	29
Turkey	73	24
Duck	57	27
Cayman	46	26
Cow	33	29
Rabbit	41	26
Leopard frog	26	28
Bullfrog	29	26
Halibut	49	29
Dogfish	42	26
Lamprey	41	24

by comparing the numbers of residues of each amino acid in a molecule; in point of fact, two entirely different proteins could have an identical amino acid composition, just as in the case of two different words, such as "similes" and "missile." It is necessary to know the primary structure of the monomeric subunits, following which the sequences in two polypeptide chains can be placed in juxtaposition and compared with respect to changes in their amino acid residues and coding triplets as has been done in Chapters 5 and 6. In the meantime, it is of interest to note that Pesce (cited by Kaplan, 1965) showed that muscle lactic dehydrogenases from various species each have a molecular weight in the neighborhood of 140,000 to 150,000 for the tetramer and are similar in their contents of phenylalanine, although considerable differences exist with respect to histidine, as shown in Table 8-3.

The malic dehydrogenases from a wide variety of species were found to be dimers and to have a molecular weight of about 67,000. These included the enzymes from several vertebrates, an insect, the horseshoe crab, two molds, and a wide range of bacteria. A second group of bacteria has malic dehydrogenases with a molecular weight of approximately 97,000, suggesting that these contain three subunits instead of the usual two.

Phosphoglucomutase

The attention of Joshi et al. (1965) was directed to this enzyme because the reaction which it catalyzes, glucose 1-phosphate $\rightleftharpoons$ glucose 6-phosphate, must occur in virtually all living forms. The reaction was shown by Najjar (1961) to pass through an intermediary stage of glucose 1,6-diphosphate. One of the phosphates is subtracted from the glucose diphosphate by esterification with a serine residue in the enzyme.

It was found by Joshi et al. (1965) that the average molecular weight of the enzymes isolated from various species was in the neighborhood of about 67,000. Considerable diversities were found in the behavior of the phosphoglucomutases from various species, including their activation by –SH reagents. The enzymes from rabbit muscle and *E. coli* showed marked differences from each other. For example, the rabbit enzyme has two active serine residues as compared with one in the *E. coli* enzyme. The enzymes from *M. lysodeikticus* and *B. cereus* show even more pronounced differences, since it is not possible to transfer phosphate to them from glucose 1,6-diphosphate. The evidence for a possible common origin of the phosphoglucomutases is incomplete.

Evolutionary Patterns in Lipid Metabolism

The lipids are a large group of diverse substances that have in common only their solubility in the so-called fat solvents, such as ether. Lipids are always found in the membranes within cells, including mitochondrial and protoplast membranes and in the endoplasmic reticulum.

The general picture of lipid biosynthesis is one of increasing complexity as one ascends the evolutionary tree, especially in the case of the steroids, which have genetic regulatory functions in animals. The

260 Biochemistry and Evolutionary Pathways

higher plants and animals may be divided roughly into two groups with respect to their synthesis and utilization of unsaturated fatty acids. This is portrayed in Fig. 8-4, condensed from Bloch (1964). The plant group desaturates stearic acid (C_{18}) to α-linolenic acid, while the animal group produces γ-linolenic acid from linoleic acid, which is obtained in the

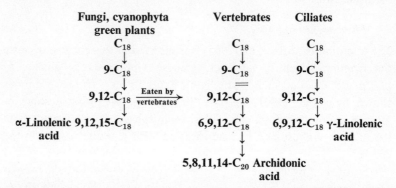

Fig. 8-4. Synthesis and utilization of unsaturated fatty acids (from Bloch, 1964). Abbreviations: C_{18} = stearic acid ($C_{17}H_{35}$ COOH); $9C_{18}$ = oleic acid (9-octadecenoic acid); $9,12C_{18}$ = linoleic acid (cis, cis-9,12-octadecadienoic acid).

diet or is produced by ciliates from stearic acid. Vertebrates desaturate γ-linolenic acid further to arachidonic acid, which is a constituent of mitochondria.

A genetic lesion has been developed by the vertebrates in this pathway. They are unable to dehydrogenate oleic acid to linoleic acid, so that they have a dietary requirement for linoleic acid. Vertebrates also have a dietary requirement for pantothenic acid owing to evolutionary loss of an enzyme that is needed for its biological synthesis. Pantothenic acid is a component of coenzyme A, which participates in the synthesis of all fats and steroids.

The biological synthesis of the terpene and sterol groups is shown in Fig. 8-5. The pathways illustrate major evolutionary divergences in a group of metabolic pathways. Once again, certain organisms have lost their synthetic abilities and have become dependent on other species; for example, some insects need cholesterol as a dietary factor,

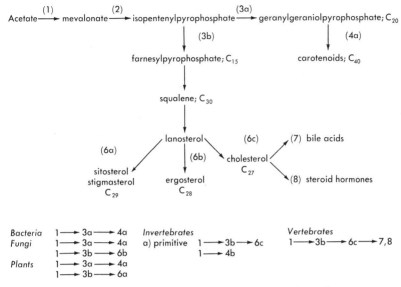

Fig. 8-5. Biosynthetic pathways of terpenes and sterols
(from Bloch, 1964).

and the vertebrates cannot synthesize carotene although they can transform it into vitamin A.

These metabolic pathways show the acquisition of new enzymes by certain groups of organisms and their loss by others as a result of changes in the genes.

The Attitude of Human Beings toward Evolution

During the past century the concept of biochemical evolution has aroused controversies with respect to its validity, especially as regards its relation to human beings. The basic disputes that followed the publication of *The Origin of Species* have been described many times. A public confrontation took place in 1860 between Thomas Huxley and Bishop Wilberforce that has since become famous. Wilberforce asked Huxley if it was from his mother's or his father's side that he claimed descent from the apes, to which Huxley retorted, in effect, that he would prefer to be descended from an ape than from the good

Bishop. The freedom of such an exchange is in refreshing contrast to the constraints placed on debate in less tolerant eras and countries.

In 1925 an amused world witnessed the attempt by the State of Tennessee to legislate against evolution as follows:

It shall be unlawful for any teacher in any of the universities, normals and all other public schools of the state which are supported in whole or in part by the public school funds of the state, to teach any theory that denies the story of the divine creation of man as taught in the Bible, and to teach instead that man has descended from a lower order of animals.

This ordinance led to the Scopes trial in Dayton, in which the defendant was fined for teaching evolution in the local high school. The successful prosecutor was William Jennings Bryan, who died two days after the verdict, allegedly following the consumption of inordinately large quantities of mashed potatoes. We shall allude to potatoes again later. Bryan did not live to see the decision reversed by the Tennessee Supreme Court, but the defense lawyer, Clarence Darrow, did and gained great fame from this and other celebrated cases. Beyond causing astonishment, the incident had no perceptible effect on research in genetics and evolution in the United States.

Grim and tragic stories came from Russia in the 1930s and 1940s, when the Communists found that science was at odds with their dogma of human genetic uniformity. Some examples of the treatment of geneticists were as follows:

B. S. Chetverikoff, W. P. Ephroimson and L. P. Ferry were banished to Siberia. The last named, together with I. J. Agol, were the first geneticists to be put to death, in 1935. In 1937, S. G. Levit met the same fate, as did (probably) Avdoulov. In 1939, N. A. Iljin joined the ranks of the martyrs; in 1942, the roster was swelled with the names of Vavilov and G. D. Karpechenko. We mention only the more distinguished men whose fate is known with a considerable degree of probability. Many others simply disappeared (Hardin, 1961, p. 205).

A strange figure now stepped to the center of the Russian scientific stage—T. D. Lysenko. His activities on behalf of fantasy crippled Soviet agriculture to an extent that made him possibly the greatest ally of the Western nations in the struggles of post-World War II decade. For example, his followers "began to argue against the very existence of plant viruses, picturing them as a metaphysical construct of bourgeois science" (Joravsky, 1962). As a result of this and similar attitudes, the yields of potatoes per acre in the Soviet Union, where they are an essential staple, dropped during 1955–58 to less than one-half of the corresponding levels in the United States where the excess seems to be used to a large extent for potato chips.

The wrath of Lysenko and his associates was directed against "Mendel-Morganism," which, in molecular terms, means against DNA. Lysenko stated, for example, that one species of wheat with 28 chromosomes could be converted into several varieties of soft 42-chromosome wheat and that the conversion of one species into another takes place by a leap. The climax of Lysenko's power was reached in 1948 at the Lenin Academy when he made the following announcement:

Comrades! Before proceeding to the concluding remarks, I consider it my duty to declare the following: I have been asked in one of the memoranda as to the attitude of the Central Committee concerning my paper. I answer: the Central Committee of the Party has examined my report and approved it (Hardin, 1961, p. 211).

The reaction to this statement was reported by *Pravda* as follows:

The communication by the President aroused general enthusiasm in the members of the session. As if moved by a single impulse, all those present arose from their seats and started a stormy, prolonged ovation in honor of the Central Committee of the Lenin-Stalin Party, in honor of the wise leader and teacher of the Soviet people, the greatest scientist of our era, Comrade Stalin (*ibid.*).

Eight years earlier, in California, Beadle and Tatum had stated the "one gene–one enzyme" principle. A few years later Watson and Crick laid bare the structure of the molecule of DNA. From these and from related discoveries there came a flood of scientific evidence for the "DNA-RNA-protein" concept that imposes such difficulties on those who advocate the doctrine of the inheritance of acquired characters. The Lysenkoists who wish to change the facts of heredity and evolution must now direct their attacks not on individuals, but on adenine, guanine, cytosine, thymine, uracil, the amino acids, ribosomes, and enzymes. These molecules are cold and unresponsive toward oratory. There are indications, however, that Mahomet will go to the mountain: in 1963 a Russian geneticist stated "that Lysenkoists still denied the existence of genes but were willing to accept the existence of DNA as hereditary material" (Caspari and Marshak, 1965).

One of the most famous incidents in biology concerns the discovery of the basic mechanism of heredity by Mendel in 1866 and its neglect by his contemporaries, following which his paper was "rediscovered" in 1900, some years after Mendel's death. Many authors have expressed their sorrow that Mendel should not have received the respect and attention that was due him in his lifetime. Yet it may well be that we should envy Mendel without reservation. Perhaps the greatest of all intellectual experiences is to be alone with Nature when she draws the veil briefly from her face and the viewer knows he has seen what no man saw before. This experience came to Gregor Mendel, and his confrontation with a new concept was interrupted by no one; he was the first to perceive the existence of the gene, and he knew what he had found. Let us hope that the desire to uncover the many hidden intricacies of life will continue to drive forward those whose imagination and zeal are awakened by the contemplation of biology.

SUMMARY

Molecular evolution stems from events that take place in DNA molecules. These events seem to be of three general types. The first of these is duplication, as distinguished from replication, of DNA double strands, either as entire chromosomes, as genes, or as segments of genes. Crossing-over and recombination may be among the principal mechanisms involved in the latter two cases. Gene duplication is characteristically followed by functional differentiation (Lewis, 1951), and this seems also to apply to segments of genes, as exemplified by the molecule of ferredoxin (p. 228). The second type of event, which may sometimes be an accompaniment of the first, is shortening, or deletion of portions of DNA strands. The third is point mutations, which often occur as replacements of one base-pair in a DNA molecule. The various changes are translated into proteins by means of the genetic code. These events apparently take place randomly, and the perpetuation or disappearance of their phenotypic consequences is decided by the classical evolutionary parameter of natural selection. The remainder of the evolutionary process is for the most part concerned with various complexities arising from morphological, ecological, and other results of the translation of the molecular changes listed above.

MAN

Five hundred million years ago, a billion years ago, the long molecules were joined together in the pools of tepid water.

Phosphate, sugar; phosphate, sugar, phosphate; A, C, T, and G; C, A, G, and T.

(What was said about the monkeys tapping on typewriters? That they could spell out all the books in the library of the British Museum? Wrong! Quite wrong! Something must read each page and pick out the good ones! Or the monkeys would write gibberish forever!)

A, C, T, and G, then C, A, G, and U, gathering the amino acids, spinning and weaving the fabrics of life, until sea-creatures left the ocean, and the warm blood flowed in their veins.

A, C, T, and G, changing to C, C, T, and G, changing to C, A, T, and G, so that he stood on two legs and hurled a stone. C, A, T, and G told him, "You must steal the fire that the lightning brought to the pine tree. Tie the sharp stone to a stick. Scratch pictures on the cliff. Bow down before the sun. Show your child all these things, so that he shall drive the wild beasts before him."

(His son's sons shall shut their eyes to the message of how our story is inscribed. Vainly they shall set themselves up as the children of gods. Until, at last, young men shall read what we have written, and tell each other whence they came.)

(Five hundred million years, a billion years, the long rods, immortally never-changing, mortally ever-changing, reached the day, when, through what they had wrought, they saw themselves as in a mirror.)

References

Aach, H. G., G. Funatsu, M. W. Nirenberg, and H. Fraenkel-Conrat. 1964. Biochemistry 3:1362.
Acs, G., E. Reich, and S. Valanja. 1963. Biochem. Biophys. Acta 76:68.
Adams, J., and M. Capecchi. 1966. Proc. Natl. Acad. Sci. U.S. 55:147.
Allan, N., D. Beale, D. Irvine, and H. Lehmann. 1965. Nature 208:658.
Allen, E., E. Glassman, E. Cordes, and R. S. Schweet. 1960. J. Biol. Chem. 235:1068.
Allison, A. C. 1954. Brit. Med. J. 1954(1):290.
────── 1964. Cold Spring Harbor Symp. Quant. Biol. 29:137.
Ambler, R. P. 1962. Biochem. J. 82:30.
────── 1963. Biochem. J. 89:349.
Ames, B. N., and B. Garry. 1959. Proc. Natl. Acad. Sci. U.S. 45:1453.
Ames, B. N., and P. E. Hartman. 1963. Cold Spring Harbor Symp. Quant. Biol. 28:349.
Anderer, F. A. 1962. Z. Naturforsch. 17b:526.
Anfinsen, C. B. 1959. The molecular basis of evolution. Wiley, New York.
Armstrong, A., H. Hagopian, V. M. Ingram, I. Sjöquist, and J. Sjöquist. 1964. Biochemistry 3:1194.
Auerbach, C., and J. M. Robson. 1946. Nature 157:302.
Avery, O. T., C. M. Macleod, and M. McCarty. 1944. J. Exptl. Med. 79:137.
Baglioni, C. 1962a. Biochim. Biophys. Acta 59:437.
────── 1962b. J. Biol. Chem. 237:69.
────── 1963. In J. H. Taylor [ed.], Molecular genetics, p. 405. Academic, New York.
────── 1965. Reported by D. Beale and H. Lehmann, Nature 207:159.
────── and V. M. Ingram. 1961. Biochem. Biophys. Acta 48:253.
Baglioni, C., and H. Lehmann. 1962. Nature 196:229.
Bahl, O. P., and E. L. Smith. 1965. J. Biol. Chem. 240:3585.
Banting, F. G., and C. H. Best. 1922. J. Lab. Clin. Med. 7:251.
Barnabas, J., and C. J. Muller. 1962. Nature 194:931.
Barnett, W. E., and K. B. Jacobson. 1964. Proc. Natl. Acad. Sci. U.S. 51:642.
Barondes, S. H., and M. W. Nirenberg. 1962. Science 138:810, 813.

Baumberg, S., D. F. Bacon, and H. Vogel. 1965. Proc. Natl. Acad. Sci. U.S. 53:1029.
Bautz, E. K. F., and B. D. Hall. 1962. Proc. Natl. Acad. Sci. U.S. 48:400.
Beadle, G. W., and E. L. Tatum. 1941. Proc. Natl. Acad. Sci. U.S. 27:499.
Beale, D., and H. Lehmann. 1965. Nature 207:159.
Belozersky, A. N., and A. S. Spirin. 1960. *In* E. Chargaff and J. N. Davidson [eds.], The nucleic acids: Chemistry and biology, III, 147. Academic, New York.
Benzer, S., and S. P. Champe. 1962. Proc. Natl. Acad. Sci. U.S. 48:1114.
Benzer, S., and E. Freese. 1958. Proc. Natl. Acad. Sci. U.S. 44:112.
Benzer, S., and B. Weisblum. 1961. Proc. Natl. Acad. Sci. U.S. 47:1149.
Benzinger, R., and P. E. Hartman. 1962. Virology 18:614.
Berg, P. 1956. J. Biol. Chem. 222:1025.
———, F. H. Bergmann, E. J. Ofengand, and M. Dieckmann. 1961. J. Biol. Chem. 236:1726.
Berg, P., U. Lagerkvist, and M. Dieckmann. 1962. J. Mol. Biol. 5:159.
Bernfield, M. R., and M. W. Nirenberg. 1965. Science 147:479.
Bessman, M. J., I. R. Lehman, J. Adler, S. B. Zimmerman, E. S. Simms, and A. Kornberg. 1958. Proc. Natl. Acad. Sci. U.S. 44:633.
Bloch, K. 1964. *In* C. A. Leone [ed.], Taxonomic biochemistry and serology, p. 728. Ronald, New York.
Bolton, E. T., and B. J. McCarthy. 1962. Proc. Natl. Acad. Sci. U.S. 48:1390.
Borek, E. 1965. The code of life. Columbia Univ. Press, New York.
Borsook, H., E. H. Fischer, and G. Keighley. 1957. J. Biol. Chem. 229:1059.
Bowman, B. H., C. P. Oliver, D. R. Barnett, J. E. Cunningham, and R. G. Schneider. 1964. Blood 23:193.
Boyce, R. P., and P. Howard-Flanders. 1964. Proc. Natl. Acad. Sci. U.S. 51:293.
Brachet, J., and A. E. Mirsky [eds.]. 1958–64. The cell: Biochemistry, physiology, morphology, 5 vols. Academic, New York.
Braunitzer, G. 1965. *In* V. Bryson and H. J. Vogel [eds.], Evolving genes and proteins, p. 183. Academic, New York.
——— and V. Braun. 1965. Z. Physiol. Chem. 340:88.
Braunitzer, G., K. Gehring-Müller, N. Hilschmann, K. Hilse, J. Hobom, V. Rudloff, and B. Wittmann-Liebold. 1961. Z. Physiol. Chem. 325:283.
Braunitzer, G., K. Hilse, V. Rudloff, and N. Hilschmann. 1964. Advan. Protein Chem. 19:1.

Brenner, S. 1957. Proc. Natl. Acad. Sci. U.S. 43:687.
———, L. Barnett, F. H. C. Crick, and A. Orgel. 1961. J. Mol. Biol. 3:121.
Brenner, S., and J. R. Beckwith. 1965. J. Mol. Biol. 13:629.
Brenner, S., F. Jacob, and M. Meselson. 1961. Nature 190:576.
Brenner, S., A. O. W. Stretton, and S. Kaplan. 1965. Nature 206:994.
Bretscher, M. S., H. M. Goodman, J. R. Menninger, and J. D. Smith. 1965. J. Mol. Biol. 14:634.
Brimacombe, R., J. Trupin, M. Nirenberg, P. Leder, M. Bernfeld, and T. Jaouni. 1965. Proc. Natl. Acad. Sci. U.S. 54:954.
Brinton, C. C., Jr., and L. S. Baron. 1960. Biochim. Biophys. Acta 42:298.
Buchanan, B. B., W. Lovenberg, and J. C. Rabinowitz. 1963. Proc. Natl. Acad. Sci. U.S. 49:345.
Buettner-Janusch, J., and R. L. Hill. In V. Bryson and M. J. Vogel [eds.], Evolving genes and proteins, p. 167. Academic, New York.
Burma, D. P., H. Kroger, S. Ochoa, R. C. Warner, and J. D. Weill. 1961. Proc. Natl. Acad. Sci. U.S. 47:749.
Cairns, J. 1963. Cold Spring Harbor Symp. Quant. Biol. 28:43.
Cantoni, G. L., H. Ishikura, H. H. Richards, and U. Tanaka. 1963. Cold Spring Harbor Symp. Quant. Biol. 28:123.
Capecchi, M. R., and G. N. Gussin. 1965. Science 149:417.
Carbon, J. A., L. Hung, and D. Jones. 1965. Federation Proc. 24:216.
Caspar, D. L. D. 1963. Advan. Protein Chem. 18:37.
Caspari, E. W., and E. Marshak. 1965. Science 149:275.
Cavalieri, L. F., J. J. Fox, A. Stone, and N. Chang. 1954. J. Am. Chem. Soc. 76:1119.
Champe, S. P., and S. Benzer. 1962*a*. Proc. Natl. Acad. Sci. U.S. 48:532.
——— 1962*b*. J. Mol. Biol. 4:288.
Chan, S. K., and E. Margoliash. 1966. J. Biol. Chem 241:335.
Chan, S. K., S. B. Needleman, J. W. Stewart, O. F. Walasek, and E. Margoliash. 1963. Federation Proc. 22:658.
Chargaff, E. 1955. In E. Chargaff and J. N. Davidson [eds.], The nucleic acids: Chemistry and biology, I, 307. Academic, New York.
Chernoff, A. K., and P. E. Perrillie. 1964. Biochem. Biophys. Res. Commun. 163:68.
Clark, B. F. C., and K. A. Marcker. 1965. Nature 207:1038.
Clegg, J. B., M. A. Naughton, and D. J. Weatherall. 1965. Nature 207:945.
Cohen, S. S. 1963. Science 139:1017.
Commoner, B. 1965. Proc. Natl. Acad. Sci. U.S. 53:1183.
Crick, F. H. C. 1958. Symp. Quant. Biol. 12:138.
——— 1965. Codon-anticodon pairing—the wobble hypothesis. [Circulated to Information Exchange Group 7. Natl. Inst. of Health, June, 1965.]

Crick, F. H. C., L. Barnett, S. Brenner, and R. J. Watts-Tobin. 1961. Nature 192:1227.

Crick, F. H. C., J. S. Griffith, and L. E. Orgel. 1957. Proc. Natl. Acad. Sci. U.S. 43:416.

Crookston, J. H., D. Beale, D. Irvine, and H. Lehmann. 1965. Nature 208:1059.

Darwin, C. 1859. The origin of species. Chapter 14. [Reprinted, Murray, London, 1929, 6th ed., p. 603.]

Davies, J., L. Gorini, and B. D. Davis. 1965. Mol. Pharmacol. 1:93.

Demerec, M. 1965. In V. Bryson and H. J. Vogel [eds.], Evolving genes and proteins, p. 505. Academic, New York.

────── and P. E. Hartman. 1959. Ann. Rev. Microbiol. 13:377.

Dickerson, R. E. 1964. In H. Neurath [ed.], The proteins, II, 603. Academic, New York, 2d ed.

Dintzis, H. M. 1961. Proc. Natl. Acad. Sci. U.S. 47:247.

Doolittle, R. F., and B. Blombäck. 1964. Nature 202:147.

Doty, P., J. Marmur, J. Eigner, and C. L. Schildkraut. 1960. Proc. Natl. Acad. Sci. U.S. 46:461.

Dounce, A. L. 1952. Enzymologia 15:251.

Drake, J. W. 1963. J. Mol. Biol. 6:268.

Dütting, D., W. Karau, F. Melchers, and H. G. Zachau. 1965. Biochim. Biophys. Acta 108:194.

Dunn, D. B. 1959. Biochim. Phys. Acta 34:286.

────── and J. D. Smith. 1954. Nature 174:305.

Dure, L., and L. Waters. 1965. Science 147:410.

Dus, K., R. G. Bartsch, and M. D. Kamen. 1962. J. Biol. Chem. 237:3083.

Eck, J., and C. S. Marvel. 1934. J. Biol. Chem. 106:387.

Eck, R. V. 1963. Science 140:477.

────── 1964. Proc. Conf. Biol. Med. 17:115.

Edgar, R. S. 1962. Ann. Rept. Calif. Inst. Technol. (Biology), p. 146.

────── and M. Susman. 1962. Ann. Rept. Calif. Inst. Technol. (Biology), p. 145.

Edmundson, A. B. 1965. Nature 205:883.

Engelhart, J. F. 1825. Dissertation, Göttingen. [Cited in Braunitzer et al. (1964).]

Epstein, C. J., and A. G. Motulsky. 1965. In A. G. Steinberg and A. G. Bearn [eds.], Progress in medical genetics, p. 85. Grune & Stratton, New York.

Epstein, R. H., A. Bolle, C. M. Steinberg, A. Kellenberger, E. Boy de la Tour, R. Chevalley, A. Edgar, M. Susman, G. H. Denhardt, and A. Lielausis. 1963. Cold Spring Harbor Symp. Quant. Biol. 28:375.

Falkow, S., R. Rownd, and L. S. Baron. 1962. J. Bacteriol. 84:1303.
Firschein, L. 1961. Am. J. Human Genet. 24:375.
Fitch, W. M. 1966. J. Mol. Biol. (in press).
Freese, E. 1959a. J. Mol. Biol. 1:87.
——— 1959b. Proc. Natl. Acad. Sci. U.S. 45:622.
——— 1963. In J. H. Taylor [ed.], Molecular genetics, p. 207. Academic New York.
Freifelder, D., A. K. Kleinschmidt, and R. L. Sinsheimer. 1964. Science 146:255.
Funatsu, G., and H. Fraenkel-Conrat. 1964. Biochemistry 3:1356.
Funke, W. 1851. Z. Rat. Med. 1:185.
Gamow, G. 1954a. Dansk Biol. Med. 22:No. 3.
——— 1954b. Nature 173:318.
———, A. Rich, and M. Yčas. 1956. Advan. Biol. Med. Phys. 4:23.
Gamow, G., and M. Yčas. 1955. Proc. Natl. Sci. U.S. 41:1011.
Garen, A., C. Levinthal, and F. Rothman. 1961a. Deoxyribonucleic acid, p. 190. Macmillan, New York.
——— 1961b. J. Chem. Phys. 58:1068.
Garen, A., and O. Siddiqi. 1962. Proc. Natl. Acad. Sci. U.S. 48:1121.
Garen, A., and N. D. Zinder. 1955. Virology 1:347.
Gatlin, L. L. 1963. J. Theoret. Biol. 5:360.
Geiduschek, E. P. 1961. Proc. Natl. Acad. Sci. U.S. 47:950.
———, J. W. Moohr, and S. B. Weiss. 1962. Proc. Natl. Acad. Sci. U.S. 48:1078.
Gerald, P. S., and M. L. Efron. 1961. Proc. Natl. Acad. Sci. U.S. 47:1758.
Giaconomi, P. S., and S. Spiegelman. 1962. Science 138:1328.
Gierer, A., 1961. Proc. 5th Intern. Biochem. Congr. (Moscow).
Gilbert, W. 1963. J. Mol. Biol. 6:374.
Gillespie, D., and S. Spiegelman. 1966. J. Mol. Biol. (in press).
Gold, M., and J. Hurwitz. 1964. J. Biol. Chem. 239:3866.
———, and M. Anders. 1963. Proc. Natl. Acad. Sci. U.S. 50:164.
Goldfine, H., and M. E. Ellis. 1964. J. Bacteriol. 87:8.
Goodman, H. M., and A. Rich. 1962. Proc. Natl. Acad. Sci. U.S. 48:2101.
Gordon, W. G., J. J. Basch, and E. B. Kalan. 1961 J. Biol. Chem. 236:2908.
Gottlieb, A. J., A. Restrepo, and H. A. Itano. 1963. Federation Proc. 23:172.
Green, M. H. 1965. Proc. Natl. Acad. Sci. U.S. 52:1388.
Griffith, W. H., and D. J. Mulford. 1941. J. Am. Chem. Soc. 63:929.
Guest, J., and C. Yanofsky. 1965. J. Mol. Biol. 12:793.

Haldane, J. B. S. 1932. The inequality of man, and other essays, p. 148, Chatto and Windus, London. [Also published as Science and human life. Harper, New York, 1933. Quoted in Horowitz (1945).]
Hall, B. D., and S. Spiegelman. 1961. Proc. Natl. Acad. Sci. U.S. 47:137.
Hall, R. H., and G. B. Chheda. 1965. J. Biol. Chem. 240:PC2754.
Hanadu, M., and D. Rucknagel. 1963. Biochem. Biophys. Res. Commun. 11:229.
Harada, K., and S. W. Fox. 1965. In S. W. Fox [ed.], The origin of prebiological systems and of their molecular matrices, p. 187. Academic, New York.
Hardin, G. 1961. Nature and man's fate, pp. 205, 211. Mentor, New York.
Hartley, B. S. 1964. Nature 201:1284.
———, J. R. Brown, D. L. Kauffman, and L. B. Smillie. 1965. Nature 207:1157.
Haselkorn, R., and V. A. Fried. 1964. Proc. Natl. Acad. Sci. U.S. 51:1001.
Hayashi, M., M. N. Hayashi, and S. Spiegelman. 1963. Proc. Natl. Acad. Sci. U.S. 50:664.
Hecht, L. I., M. L. Stephenson, and P. C. Zamecnik. 1959. Proc. Natl. Acad. Sci. U.S. 45:505.
Heller, J., and E. L. Smith. 1965. Proc. Natl. Acad. Sci. U.S. 54:1621.
Henning, U., D. R. Helinski, F. C. Chao, and C. A. Yanofsky. 1962. J. Biol. Chem. 237:1523.
Heppel, L. A., P. J. Ortiz, and S. Ochoa. 1957. J. Biol. Chem. 229:695.
Heppel, L. A., P. R. Whitfield, and R. Markham. 1955. Biochem. J. 60:8.
Hill, R. L. 1964. [Personal communication.]
———, J. Buettner-Janusch, and V. Buettner-Janusch. 1963. Proc. Natl. Acad. Sci. U.S. 50:885.
Hill, R. L., and H. C. Schwartz. 1959. Nature 184:641.
Hiraizumi, Y., L. Sandler, and J. F. Crow. 1960. Evolution 14:433.
Hitzig, W. H., P. G. Frick, K. Betke, and T. H. Huisman. 1962. Helv. Paediat. Acta 15:499.
Hoagland, M. B. 1955. Biochim. Biophys. Acta 16:288.
———, E. B. Keller, and P. C. Zamecnik. 1956. J. Biol. Chem. 218:345.
Holley, R. W., J. Apgar, G. A. Everett, J. T. Madison, M. Marquisee, S. H. Merrill, J. R. Penswick, and A. Zamir. 1965. Science 147:1462.
Hoppe-Seyler, R. 1862. Arch. Path. Anat. Physiol. 23:446.
——— 1864. Arch. Path. Anat. Physiol. 29:233, 567.
Horowitz, N. H. 1945. Proc. Natl. Acad. Sci. U.S. 31:153.
——— 1965. In V. Bryson and H. J. Vogel [eds.], Evolving genes and proteins, p. 15. Academic, New York.

Howard, B. D., and I. Tessman 1964a. J. Mol. Biol. 9:364.
——— 1964b. J. Mol. Biol. 9:372.
Hoyer, B. H., E. T. Bolton, B. J. McCarthy, and R. B. Roberts. 1965. *In* V. Bryson and H. J. Vogel [eds.], Evolving genes and proteins, p. 581. Academic, New York.
Hoyer, B. H., B. J. McCarthy, and E. T. Bolton. 1963. Science 140:1408.
——— 1964. Science 144:959.
Huehns, E. R., and E. M. Shooter. 1965. J. Med. Genetics 2:48.
Huisman, T. H., K. Punt, and D. D. G. Schaad. 1961. Blood 17:747.
Hunt, J. A., and V. M. Ingram. 1959a. Nature 184:640.
——— 1959b. Nature 184:870.
Hurwitz, J., M. Anders, M. Gold, and I. Smith. 1965. J. Biol. Chem. 240:1256.
Ingram, V. M. 1957. Nature 180:326.
——— 1958. Sci. American 198:68.
——— 1959. Biochim. Biophys. Acta 36:402.
——— 1962. *In* J. M. Allen [ed.], The molecular control of cellular activity p. 179. McGraw-Hill, New York.
——— 1963. The hemoglobins in genetics and evolution. Columbia Univ. Press, New York.
Itano, H. A. 1960. *In* Genetics, p. 136. Josiah Macy, Jr., Foundation Symposium, New York.
Jacob, F., and J. Monod. 1961. J. Mol. Biol. 3:318.
Jacob, F., D. Perrin, C. Sanchez, and J. Monod. 1960. Compt. Rend. 250:1727.
Jacob, F., and E. L. Wollman. 1957. Compt. Rend. 245:1840.
Jones, R. T., F. W. Boerma, and T. H. J. Huisman. 1965. Am. J. Human Genet. 17:511.
Jones, R. T., B. Brimhall, E. R. Huehns, and N. A. Barnicot. 1966. Science 151:1406.
Jones, R. T., B. Brimhall, and T. H. J. Huisman. 1966. Clin. Res. (in press).
Jones, R. T., R. B. Coleman, and P. Heller. 1964. Federation Proc. 23:173.
Joravsky, D. 1962. Sci. American 207:41.
Joshi, J. G., T. Hashimoto, K. Hanabusa, H. W. Dougherty, and P. Handler. 1965. *In* V. Bryson and H. J. Vogel [eds.], Evolving genes and proteins, p. 207. Academic, New York.
Josse, J., A. D. Kaiser, and A. Kornberg. 1961. J. Biol. Chem. 236:864.
Jukes, T. H. 1963a. Advan. Biol. Med. Phys. 9:1.
——— 1963b. Biochem. Biophys. Res. Commun. 10:155.

Jukes, T. H. 1965a. Biochem. Biophys. Res. Commun. 19:391.
───── 1965b. *In* S. W. Fox [ed.], The origins of prebiological systems and of their molecular matrices, p. 407. Academic, New York.
Kalan, E. B., R. Greenberg, M. Walter, and W. C. Gordon. 1964. Biochem. Biophys. Res. Commun. 16:199.
Kaplan, N. O. 1965. *In* V. Bryson and H. J. Vogel [eds.], Evolving genes and proteins, p. 243. Academic, New York.
Kaplan, S., A. O. W. Stretton, and S. Brenner. 1965. J. Mol. Biol. 14:528.
Kaziro, Y., A. Grossman, and S. Ochoa. 1963. Proc. Natl. Acad. Sci. U.S. 50:54.
Keller, E. B., and R. S. Anthony. 1963. Federation Proc. 22:231.
Kendrew, J. C. 1962. Brookhaven Symp. Biol. 15:216.
Khorana, H. G. 1965. Federation Proc. 24:1473.
Knight, C. A. 1963. *In* Handbuch der Protoplasmaforschung, IV, 59. Springer, Vienna.
Konigsberg, R., G. Notani, and N. D. Zinder. 1966. [In preparation.]
Konigsberg, W., J. Goldstein, and R. J. Hill. 1963. J. Biol. Chem. 238:2028.
Kotaki, A. 1963. J. Biochem. 53:61.
───── and S. Ochoa. 1963. Proc. Natl. Acad. Sci. U.S. 49:88.
Kreil, G. 1963. Z. Physiol. Chem. 334:154.
Kraus, L. 1965. Reported by D. Beale and H. Lehmann. Nature 207:159.
Kruh, J., and H. Borsook. 1956. J. Biol. Chem. 220:905.
Lawley, P. D., and P. Brooks. 1962. J. Mol. Biol. 4:216.
Lazzarini, R. A., and A. Peterkofsky. 1965. Proc. Natl. Acad. Sci. U.S. 53:549.
Leder, P., and M. W. Nirenberg. 1964a. Proc. Natl. Acad. Sci. U.S. 52:420.
───── 1964b. Proc. Natl. Acad. Sci. U.S. 52:1521.
Leder, P., F. Rottman, R. Brimacombe, J. Trupin, C. O'Neal, and M. W. Nirenberg. 1965. Federation Proc. 24:408.
Lehmann, H., D. Beale, and F. S. Bio-Daku. 1964. Nature 203:363.
Lengyel, P., J. F. Speyer, and S. Ochoa. 1961. Proc. Natl. Acad. Sci. U.S. 47:1936.
Leone, C. A. [ed.] 1964. Taxonomic biochemistry and serology, Chaps. 5, 7, 15–17, 29–37, 41–47. Ronald, New York.
Levinthal, C., E. R. Signer, and K. Fetherolf. 1962. Proc. Natl, Acad. Sci. U.S. 48:1220.
Lewis, E. B. 1951. Cold Spring Harbor Symp. Quant. Biol. 16:159.
Li, C. H., L. Barnafi, M. Chretien, and D. Chung. 1965. Nature 208:1093.
Lipsett, M. N. 1965. Biochem. Biophys. Res. Commun. 20:224.
Lisker, R. G., J. V. Ruiz-Reyes, and A. Loria. 1963. Blood 22:342.

Litman, R. M., and A. B. Pardee. 1956. Nature 178:529.
Loper, J. C., M. Grabnar, R. C. Stahl, Z. Hartman, and P. E. Hartman. 1964. Brookhaven Symp. Biol. 17:15.
Luria, S. E. 1962. Ann. Rev. Microbiol. 16:205.
McCarthy, B. J., and E. T. Bolton. 1963. Proc. Natl. Acad. Sci. U.S. 50:156.
—— 1964. J. Mol. Biol. 8:184.
McCarthy, B. J., and B. H. Hoyer. 1964. Proc. Natl. Acad. Sci. U.S. 52:915.
McDowell, M. A., and E. L. Smith. 1965. J. Biol. Chem., 240:4635.
Maitra, U., and J. Hurwitz. 1965. Proc. Natl. Acad. Sci. U.S. 54:815.
Mandel, L. R., and E. Borek. 1963. Biochemistry 2:560.
Manski, W., T. P. Auerbach, and S. P. Halbert. 1960. Am. J. Opthalmol. 50:985.
Manski, W., S. P. Halbert, and T. P. Auerbach. 1961. Arch. Biochem. Biophys. 92:512.
Marchis-Mouren, G., and F. Lipmann. 1965. Biochemistry 53:1147.
Marcker, K., and F. Sanger. 1964. J. Mol. Biol. 8:835.
Margoliash, E. 1963. Proc. Natl. Acad. Sci. U.S. 50:672.
—— and E. L. Smith. 1965. *In* V. Bryson and H. J. Vogel [eds.], Evolving genes and proteins, p. 221. Academic, New York.
——, G. Kreil, and H. Tuppy. 1961. Nature 192:1125.
Marmur, J., and P. Doty. 1961. J. Mol. Biol. 3:585.
Marmur, J., S. Falkow, and M. Mandel. 1963*a*. Ann. Rev. Microbiol. 17:329.
Marmur, J., and C. M. Greenspan. 1963. Science 142:387.
Marmur, J., and D. Lane. 1960. Proc. Natl. Acad. Sci. U.S. 46:453.
Marmur, J., C. L. Schildkraut, and P. Doty. 1961. *In* The molecular basis of neoplasia. Univ. of Texas Press, Austin.
Marmur, J., E. Seaman, and J. Levine. 1963*b*. J. Bacteriol. 85:461.
Martin, R. G. 1963. Cold Spring Harbor, Symp. Quant. Biol. 28:357.
Matsubara, H., and E. L. Smith. 1963. J. Biol. Chem. 238:2732.
Meselson, M., and F. W. Stahl. 1958. Proc. Natl. Acad. Sci. U.S. 44:671.
Miller, S. L. 1955. J. Am. Chem. Soc. 77:2351.
Montagnier, L., and F. K. Sanders. 1963. Nature 199:664.
Mortenson, L. E., R. C. Valentine, and J. E. Carnahan. 1962. Biochem. Biophys. Res. Commun. 7:448.
Moss, B., and V. M. Ingram. 1965. Proc. Natl. Acad. Sci. U.S. 54:967.
Muller, C. J., and S. Kingma. 1961. Biochem. Biophys. Acta 50:595.
Muller, H. J. 1922. Am. Naturalist 56:32.

Muller, H. J. 1927. Science 66:84.
Mundry, K. W., and A. Gierer. 1958. Z. Vererbungslehre 89:614.
Munkres, K. D., and F. M. Richards. 1965. Arch. Biochem. Biophys. 109:457.
Murayama, M. 1957. J. Biol. Chem. 228:231.
——— 1962. Nature 194:933.
——— 1965a. Federation Proc. 24:533.
——— 1965b. [Personal communication.]
Murray, R. G. E. 1962. Symp. Soc. Gen. Microbiol. 12:119.
Najjar, V. A. 1961. In The enzymes, VI, 161. Academic, New York, 2d ed.
Nakajima, H., T. Takemura, O. Nakajima, and K. Yamaoka. 1963. J. Biol. Chem. 238:3784.
Nakamoto, T., and S. B. Weiss. 1962. Proc. Natl. Acad. Sci. U.S. 48:880.
Nance, W. E. 1963. Science 141:123.
——— and O. Smithies. 1963. Nature 198:869.
Narita, K., and K. Titani. 1965. Proc. Japan. Acad. 41:831.
———, Y. Yaoi, and H. Murakami. 1963. Biochem. Biophys. Acta 77:688.
Naughton, M. A., and H. M. Dintzis. 1962. Proc. Natl. Acad. Sci. U.S. 48:1822.
Neel, J. V., I. C. Wells, and H. A. Itano. 1951. J. Clin. Invest. 30:1120.
Nirenberg, M. W., O. W. Jones, P. Leder, B. F. C. Clark, W. S. Sly, and S. Pestka. 1963. Cold Spring Harbor Symp. Quant. Biol. 28:549.
Nirenberg, M. W., and P. Leder. 1964. Science 145:1399.
———, M. Bernfield, R. L. C. Brimacombe, J. S. Trupin, F. M. Rottman, and C. O'Neal. 1965. Proc. Natl. Acad. Sci. U.S. 53:1161.
Nirenberg, M. W., and J. H. Matthaei. 1961. Proc. Natl. Acad. Sci. U.S. 47:1588.
Nishimura, S., D. S. Jones, R. D. Wells, T. M. Jacob, and H. G. Khorana. 1965. Federation Proc. 24:409.
Nomura, M., B. D. Hall, and S. Spiegelman. 1960. J. Mol. Biol. 2:306.
——— 1961. J. Mol. Biol. 3:318.
Notani, G. W., D. L. Engelhardt, W. Königsberg, and N. Zinder. 1965. J. Mol. Biol. 12:439.
Nuttall, G. H. F. 1904. Blood immunity and blood relationship. Macmillan, New York.
Nygaard, A. P., and B. D. Hall. 1963. Biochem. Biophys. Res. Commun. 12:98.
Oishi, M., and N. Sueoka. 1965. Proc. Natl. Acad. Sci. U.S. 54:483.
Okamoto, T., and M. Takanami. 1963. Biochem. Biophys. Acta 76:266.

Oparin, A. I. 1964. Life: Its nature, origin, and development, pp. 52, 65. Academic, New York.
Oró, J. 1963. Federation Proc. 22:681.
―――― 1965. In S. W. Fox [ed.], The origin of prebiological systems and of their molecular matrices, p. 137. Academic, New York.
Parker, W. C., and A. G. Bearns. 1961. Ann. Human Genet. 2:708.
Pauling, L. 1955. Harvey Lectures, Ser. 49 (1953–54):216.
―――― and R. B. Corey. 1961. Proc. Natl. Acad. Sci. U.S. 47:811.
Pauling, L., H. A. Itano, S. J. Singer, and I. C. Wells. 1949. Science 110:543.
Perutz, M. F. 1962. Nature 194:914.
―――― 1965. J. Mol. Biol. 13:646.
――――, J. C. Kendrew, and H. C. Watson. 1965. J. Mol. Biol. 13:669.
Pesce, A. 1965. Reported by N. O. Kaplan in V. Bryson and H. J. Vogel [eds.], Evolving genes and proteins, p. 243. Academic, New York.
Pierre, L. E., C. E. Rath, and K. McCoy. 1963. New Engl. J. Med. 268:862.
Pittard, J., J. S. Loutit, and E. A. Adelberg. 1963. J. Bacteriol. 85:1402.
Pittard, J., and T. Ramakrishnan. 1964. J. Bacteriol. 88:367.
Ponnamperuma, C. 1965. In S. W. Fox [ed.], The origin of prebiological systems and of their molecular matrices, p. 221. Academic, New York.
―――― and R. Mack. 1965. Science 148:1221.
Preyer, J. 1871. Die Blutkrystalle. Jena.
Rapoport, J. A. 1946. Dokl. Akad. Nauk SSSR 54:65.
―――― 1947. Dokl. Akad. Nauk SSSR 56:12.
Reich, E., G. Acs, B. Mach, and E. L. Tatum. 1962. In Informational macromolecules, p. 317. Academic, New York.
Reich, E., and I. H. Goldberg. 1964. In J. N. Davidson and W. E. Cohn [eds.], Progress in nucleic acid research and molecular biology, III, 184. Academic, New York.
Reichert, E. T., and A. P. Brown. 1907. Proc. Soc. Exptl. Biol. Med. 5:66.
―――― 1909. The differentiation and specificity of corresponding proteins and other vital substances in relation to biological classification and organic evolution: The crystallography of hemoglobins. Carnegie Institution, Washington, D.C.
Reichert, K. E. 1849. Arch. Anat. Physiol. Med., p. 198.
Rendi, R., and S. Ochoa. 1961. Science 133:1367.
Ritossa, F. M., and S. Spiegelman. 1965. Proc. Natl. Acad. Sci. U.S. 53:737.
Rogers, A. O. 1951. U.S. Patent 2,564,649.
―――― 1952. Chem. Abstr. 46:1034.

Rothfus, J. A., and E. L. Smith. 1965. J. Biol. Chem. 240:4277.
Rudner, R., H. S. Shapiro, and E. Chargaff. 1962. Nature 195:143.
Rumen, N. K., and W. E. Love. 1963. Arch. Biochem. Biophys. 103:24.
Salas, M., M. A. Smith, W. M. Stanley, Jr., A. J. Wahba, and S. Ochoa. 1965. J. Biol. Chem. 240:3988.
Sanger, F. 1952. Advan. Protein Chem. 7:1.
Sarabhai, A. S., A. O. W. Stretton, S. Brenner, and A. Bolle. 1964. Nature 201:13.
Savage, J. M. 1963. Evolution, p. 33. Holt, Rinehart and Winston, New York.
Schildkraut, C. L., K. L. Wierzchsowski, J. Marmur, D. M. Green, and P. Doty. 1962. Virology 18:43.
Schneider, R. G., and R. T. Jones. 1965. Science 148:240.
Schroeder, W. A. 1963. Ann. Rev. Biochem. 32:301.
―――, R. T. Jones, J. R. Shelton, J. B. Shelton, J. Cornick, and K. McCalla. 1961. Proc. Natl. Acad. Sci. U.S. 47:811.
Schuster, H., and G. Schramm. 1958. Z. Naturforsch. 17b:526.
Schweet, R., H. Lamfrom, and E. Allen. 1958. Proc. Natl. Acad. Sci. U.S. 44:1029.
Scott, R. B., and E. Bell. 1964. Science 145:711.
Setlow, R. B., and W. L. Carrier. 1964. Proc. Natl. Acad. Sci. U.S. 51:226.
Shibata, S., I. Iuchi, T. Miyaji, and I. Takeda. 1963. Bull. Yamaguchi Med, School 10:1.
Shibata, S., T. Miyaji, I. Iuchi, S. Ueda, and I. Takeda. 1964. Clin. Chim. Acta 10:101.
Simpson, G. G. 1953. The major features of evolution. Columbia Univ. Press, New York.
――― 1964. Science 146:1535.
Smith, D. B. 1964. Can. J. Biochem. 42:755.
Smith, M. A., M. Salas, W. M. Stanley, Jr., and A. J. Wahba. 1965. Federation Proc. 24:409.
Smithies, O. 1964. Cold Spring Harbor Symp. Quant. Biol. 29:309.
――― 1965. Proc. Prook Lodge Conf. on Proteins and Polypeptides.
―――, G. E. Connell, and G. H. Dixon. 1961. Proc. 2nd Intern. Congr. Human Genetics 2:708.
――― 1962. Nature 196:232.
Söll, D., E. Ohtsuka, D. S. Jones, R. Lohrmann, H. Hayatsu, S. Nishimura, and H. G. Khorana. 1965. Proc. Natl. Acad. Sci. U.S. 54:1378.
Sonneborn, T. M. 1965. *In* V. Bryson and H. J. Vogel [eds.], Evolving genes and proteins, p. 377. Academic, New York.

Spencer, J. H., and E. Chargaff. 1962. Biochim. Biophys. Acta 51–209.
Spencer, M., W. Fuller, M. H. F. Wilkins, and G. L. Brown. 1962. Nature 194:1014.
Spetner, L. M. 1964. J. Theoret. Biol. 7:412.
Speyer, J. F. 1965. Biochem. Biophys. Res. Comm. 21:6.
———, P. Lengyel, C. Basilio, and S. Ochoa. 1962. Proc. Natl. Acad. Sci. U.S. 48:441.
Speyer, J. F., P. Lengyel, C. Basilio, A. J. Wahba, R. S. Gardner, and S. Ochoa. 1963. Cold Spring Harbor Symp. Quant. Biol. 28:559.
Spiegelman, S., and S. A. Yankofsky 1965. *In* V. Bryson and H. J. Vogel [eds.], Evolving genes and proteins, p. 537. Academic, New York.
Spyrides, G. J., and F. Lipmann. 1962. Proc. Natl. Acad. Sci. U.S. 48:1977.
Srinivasan, P. R., and E. Borek. 1963. Proc. Natl. Acad. Sci. U.S. 49:529.
Stanier, R. Y. 1964. *In* I. C. Gunsalus and R. Y. Stanier [eds.], The bacteria: A treatise on structure and function, V, 445. Academic, New York.
——— and C. B. van Niel. 1962. Arch. Mikrobiol. 42:17.
Stanley, W. M. 1935. Science 81:644.
Starr, J. 1963. Biochem. Biophys. Res. Commun. 10:181.
Stent, G. S. 1963. Molecular biology of bacterial viruses. Freeman, San Francisco.
Storer, T. I., and R. L. Usinger. 1965. General zoology, p. 243. McGraw-Hill, New York, 4th ed.
Stretton, A. O. W., and S. Brenner. 1965. J. Mol. Biol. 12:456.
Sturtevant, A. H., and G. W. Beadle. 1962. An introduction to genetics. Dover, New York.
Sueoka, N. 1961*a*. J. Mol. Biol. 3:31.
——— 1961*b*. Cold Spring Harbor Symp. Quant. Biol. 26:35.
——— 1961*c*. Proc. Natl. Acad. Sci. U.S. 47:1141.
——— 1965. *In* V. Bryson and H. J. Vogel [eds.], Evolving genes and proteins, p. 479. Academic, New York.
———, and T. Kano-Sueoka. 1964. Proc. Natl. Acad. Sci. U.S. 52:1535.
Sueoka, N., J. Marmur, and P. Doty. 1959. Nature 183:1429.
Svensson, I., H. G. Borman, K. G. Erkisson, and K. Kjellin. 1963. J. Mol. Biol. 7:254.
Swartz, M. N., T. A. Trautner, and A. Kornberg. 1962. J. Biol. Chem. 237:1961.
Swenson, R. T., R. L. Hill, H. Lehmann, and R. T. S. Jim. 1962. J. Biol. Chem. 237:1517.

Tagawa, K., and D. I. Arnon. 1962. Nature 195:537.
Takanami, M., and T. Okamoto. 1963. J. Mol. Biol. 7:323.
Takanami, M., and Y. Yan. 1965. Proc. Natl. Acad. Sci. 54:1450.
———, and T. H. Jukes. 1965. J. Mol. Biol. 12:761.
Takanami, M., and G. Zubay. 1964. Proc. Natl. Acad. Sci. U.S. 51:834.
Tamm, C., M. E. Hodes, and E. Chargaff. 1952. J. Biol. Chem. 195:49.
Tanaka, M., T. Nakashima, A. Benson, H. F. Mower, and K. T. Yasunobu. 1964. Biochem. Biophys. Res. Commun. 16:422.
Terzaghi, E., Y. Okuda, G. Streisinger, H. Inouye, and A. Tsugita. 1965. [Personal communication.]
Tessman, I., R. K. Poddar, and J. Kumar. 1964. J. Mol. Biol. 9:352.
Thach, R. E., and T. A. Sundararajan. 1965. Proc. Natl. Acad. Sci. U.S. 53:1021.
———, and P. Doty. 1965. Federation Proc. 24:409.
Tinoco, I., Jr. The Vortex 26:318.
Tocchini-Valentini, G. P., M. Stodolsky, A. Aurisicchio, M. Sarnat, F. Graziosi, S. B. Weiss, and E. P. Geiduschek. 1963. Proc. Natl. Acad. Sci. U.S. 50:935.
Trupin, J. S., F. M. Rottman, R. L. C. Brimacombe, P. Leder, M. R. Bernfield, and M. W. Nirenberg. 1965. Proc. Natl. Acad. Sci. U.S. 53:807.
Tsugita, A. 1962a. J. Mol. Biol. 5:284.
——— 1962b. Cited by Speyer et al. (1962). Proc. Natl. Acad. Sci. U.S. 48:441.
——— and H. Fraenkel-Conrat. 1960. Proc. Natl. Acad. Sci. U.S. 46:636.
——— 1962. J. Mol. Biol. 4:73.
Van Niel, C. B. 1949. Bacteriol. Rev. 13:161.
Vaughan, M., and D. Steinberg. 1959. Advan. Protein. Chem. 14:115.
Vielmetter, W., and H. Schuster. 1960. Z. Naturforsch. 15b:304.
Vogel, H. J. 1965. *In* Evolving genes and proteins, p. 25. Academic, New York.
———, D. F. Bacon, and A. Baich. 1963. *In* Informational macromolecules, p. 293. Academic, New York.
Volkin, E., and L. Astrachan. 1956. Virology 2:149.
Wahba, A. J., C. Basilio, J. F. Speyer, P. Lengyel, R. S. Miller, and S. Ochoa. 1963. Proc. Natl. Acad. Sci. U.S. 48:1683.
Wald, G. 1961. *In* W. McElroy and B. Glass [eds.], Life and light, p. 724. Johns Hopkins Press, Baltimore.
——— 1963. Evolutionary biochemistry, p. 12. Pergamon, New York.
——— 1964. Proc. Natl. Acad. Sci. U.S. 52:595.
Waller, J. P. 1963. J. Mol. Biol. 7:483.

Waller, and J. I. Harris. 1961. Proc. Natl. Acad. Sci. U.S. 47:18.
Walsh, K. A., and H. Neurath. 1964. Proc. Natl. Acad. Sci. U.S. 52:595.
Watson, J. D., and F. H. C. Crick. 1953a. Nature 171:737.
——— 1953b. Nature 171:964.
——— 1953c. Cold Spring Harbor Symp. Quant. Biol. 18:193.
Watson-Williams, E. S., D. Beale, D. Irvine, and H. Lehmann. 1965. Nature 205:1273.
Webster, W., D. Engelhart, and N. Zinder. 1966. Proc. Natl. Acad. Sci. U.S. 55:155.
Weigert, M., E. Lanka, and A. Garen. 1965. J. Mol. Biol. 14:522.
Weigert, M. G., and A. Garen. 1965a. Nature 206:992.
——— 1965b. J. Mol. Biol. 12:448.
Weisblum, B., S. Benzer, and R. W. Holley. 1962. Proc. Natl. Acad. Sci. U.S. 48:1449.
Weismann, A. 1892. Essays upon heredity and kindred biological problems, I, 446. Oxford Univ. Press, London.
Weissmann, C., and P. Borst. 1963. Science 142:1188.
Wilkins, M. H. F., A. R. Stokes, and H. R. Wilson. 1953. Nature 171:738.
Wilt, F. H. 1959. Develop. Biol. 1:199.
Winterhalter, K. H., and E. R. Huehns. 1964. J. Biol. Chem. 239:3699.
Wittmann, H. G. 1962. Naturwissenschaften 48:729.
——— and B. Wittmann-Liebold. 1963. Cold Spring Harbor Symp. Quant. Biol. 28:589.
Wittmann-Liebold, B., and H. G. Wittmann. 1966. Z. Vererbungslehre 98 (in press).
Wittmann-Liebold, B., J. Jauregai-Adell, and H. G. Wittmann. 1965. Z. Naturforsch. 20b:1235.
Woese, C. R. 1961. Biochem. Biophys. Res. Commun. 5:88.
Wyman, J. 1963. Cold Spring Harbor Symp. Quant. Biol. 28:483.
Yamane, T., and N. Sueoka. 1963. Proc. Natl. Acad. Sci. U.S. 50:1093.
——— 1964. Proc. Natl. Acad. Sci. U.S. 51:1178.
Yankofsky, S. A., and S. Spiegelman. 1962a. Proc. Natl. Acad. Sci. U.S. 48:1069, 146.
——— 1962b. Proc. Natl. Acad. Sci. U.S. 48:1069.
——— 1963. Proc. Natl. Acad. Sci. U.S. 49:538.
Yanofsky, C. A. 1963a. Cold Spring Harbor Symp. Quant. Biol. 28:581.
——— 1963b. In Informational macromolecules, p. 195. Academic, New York.
——— 1964. In I. C. Gunsalus and R. Y. Stanier [eds.], The bacteria: A treatise on structure and function, V, 373. Academic, New York.

Yanofsky, C. A. 1965. Biochem. Biophys, Res. Commun. 18:898.
———, B. C. Carlton, J. R. Guest, D. R. Helinski, and U. Henning. 1964. Proc. Natl. Acad. Sci. U.S. 51:266.
Yanofsky, C. A., E. Cox, and V. Horn. 1966. Proc. Natl. Acad. Sci. U.S. 55:274.
Yanofsky, C. A., and D. R. Helinski. 1962. Proc. Natl. Acad. Sci. U.S. 48:173.
Yasunobu, K. T., T. Nakashima, H. Higa, H. Matsubara, and A. Benson. 1963. Biochim, Biophys. Acta 78:791.
Yčas, M. 1958. *In* H. P. Yockey [ed.], Symposium on information theory in biology, p. 70. Pergamon, New York.
Yoshikawa, H. 1965. Proc. Natl. Acad. Sci. U.S. 53:1476.
——— and N. Sueoka. 1963a. Proc. Natl. Acad. Sci. U.S. 49:559.
——— 1963b. Proc. Natl. Acad. Sci. U.S. 49:806.
Yu, C., and F. W. Allen. 1959. Biochim. Biophys. Acta 32:393.
Zamir, A., R. W. Holley, and M. Marquisee. 1965. J. Biol. Chem. 240:1267.
Zinder, N. D. 1963. *In* Informational macromolecules, p. 229. Academic, New York.
Zuckerkandl, E. 1964. J. Mol. Biol. 8:128.
——— and L. Pauling. 1962. *In* M. Kasha and B. Pullman [eds.], Horizons in biochemistry, p. 189. Academic, New York.
——— 1965. *In* V. Bryson and H. J. Vogel [eds.], Evolving genes and proteins, p. 97. Academic, New York.

Index

A chain, insulin, 14
Acridines, 103, 110; mutants, 110, 144
Actinomycin D, 36
Activating enzyme (aminoacyl-sRNA synthetase), 24, 27, 40, 77, 78; alanine, 29, 70–71
Adaptor modification, 80
Agar columns for DNA separation, 234–40
Alkaline phosphatase, 60, 106, 121–23
Amber mutations, 74, 76, 117–23
Amber triplet, 119–20, 124
Amino acids: table, 11–12; incorporation, 47–53, 62
Aminoacyl-sRNA synthetase, see Activating enzyme
Aminoadipic pathway, 247–50
2-Aminopurine, 107–9, 113, 114, 119, 136
Analysis, end-group 16–18
Anemia, sickle-cell, 125, 164–68, 184
Annealing of DNA, 231, 237
Antibiotics, 246
Anti-codon, 25, 71–73
Anti-codon deaminase, 73
Apurinic acid, 30
Arginine, biosynthetic pathway, 98, 246
Asparagine, biosynthetic pathway, 65
Azotobacter vinelandii, 45

Bacillus cereus, 32–33, 36, 259
Bacillus megaterium, 38, 92
Bacillus subtilis, 36, 39, 82–85, 92, 241
Bacteriophages, 76, 79, 102; *fd*, 70; *T4*, 60, 76, 103, 114–15, 143–44; *T2*, 80
β-Ionone, 251
β-Lactoglobulin, 128, 130

Biosynthesis: asparagine, 65; glutamine, 65; methionine, 65, 69; tyrosine, 65; tryptophan, 65, 70; choline 69, 246; arginine, 98, 246; lysine, 246–50; Vitamin A, 250–55; lipid, 259–60
Bromodeoxyuridine, 110, 113, 136, 232
Bromouracil, 103, 107, 109, 232; mutants, 110, 114
Brucella abortus, 38
Bryan, William Jennings, 262

Carotene, 250, 260
Chain-terminating triplets, 61, 115–17, 123, 128
Chain termination, 73, 121
Chloramphenicol, 85
Choline, biological synthesis, 69, 246
Chironomus thummi, hemoglobin in, 177
Chromatium, protein, 193–94; ferredoxin 227, 229
Chromosome, bacterial, 82–85
Chymotrypsinogen, 18, 127, 209–15
Cistrons, 79, 111, 116, 118
Classification by base composition, 240–42
Clostridium pasteurianum, ferredoxin, 127, 227, 229
Code: genetic, 11–74; "initiation," 88; overlapping, 23; triplet, origin of, 65
Code-altering evolutionary changes, 179
Coordinate repression, 96
Corticotropins, 220, 221
Crossing-over, 220–26
Crystallography in taxonomy, 147–49
Cytochromes c, 9, 30, 73, 109, 191–205, 249

Darrow, Clarence, 262
Deaminations by mutagens, 103, 140
Dehydrogenases, 250–59
Deletion mutation, 116, 144
DNA, 13, 14; denaturation, 34–35; polymerase, 84; homology, 230–44; annealing, 231; glucosylated, 233; separation, 234–40
DNA-like RNA, 34
Deoxyribonuclease, 30, 82
d'Herelle bodies, 102
Diaminopimelic acid pathway, 97, 247–50
Dimers, thymine, 84–85
Drosophila melanogaster, 101, 107, 127, 220, 233

Elastase, 127, 210–13
End-group analysis, 16–18
Enzyme, activating, *see* Activating enzyme
Enzyme, methylating, 43–44; lysozyme-like, 218
Escherichia coli, 34–35, 37–38, 40, 44–45, 54, 59–60, 70, 74, 78–93, 95, 98–99, 107–9, 116–18, 121–23, 133–34, 232, 239
Ethyl methane sulfonate, 108, 111–12, 114, 122, 136
Evolutionary pressures, 84

Feedback inhibition, 98
Ferredoxins, 127, 227–29
Fibrinogen, 206–9
"Fingerprinting," defined, 18
5-Fluorouracil, 108, 118
Formylmethionine, 88

Gaps, 73, 77, 132, 208; *see also* Chain-terminating triplets
Genes, 6; blocking, 36; clusters, 75, 82, 92–99; crossing-over, 220, 222, 224–27; duplication, 126; responsible for hemoglobins, 54; evolution of hemoglobin, 154–65
Germ plasm, 6
Glutamine, biosynthetic pathway, 65

Haptoglobins, 220–26; Johnson, 225
Hemoglobins, 146–90; amino acid sequences in, 156–64; crystallography, 147–49; adult A, 150, 168; gorilla, 9, 154; human, 54, 168; lamprey, 150, 165, 180, 185; amino acid interchanges in, 172–73, 178; variants, 181–84, 188; Lepore, 226; S, 164–68; synthesis, 22–23; mutational changes in, 183
Histidine, 64, 68–70, 176, 258
Histidon (histidine operon), 93–97; defined, 95
Homology and DNA, 230–44
Huxley, Thomas H., 261
Hybrid strands, 8, 35, 81, 175, 230–40
Hydroxylamine, 64, 103, 108, 111–12, 118
Hydroxyproline, 11

Incorporation errors, 108
Insulin, 14, 19, 151, 218–20
β-Ionone, 251

Lactobacillus arabinosus, 59
β Lactoglobulin, 128, 130
Lamprey, 150, 153, 165, 180, 185
Lepore hemoglobin, 226
Lipid biosynthesis, 259–60
Lysenko, T. D., 262–63
Lysine, biological synthesis, 246–50
Lysozyme, in bacteriophage *T4*, 143–45

Malaria, cerebral, 184
Meiosis, 126–27
Meiotic drive, 126–27
Mendel, Gregor, 264
"Mendel-Morganism," 263
Messenger RNA, 23, 26, 27
Metamorphosis, 250–55
Methionine, biosynthetic pathway, 65, 69
Methylating enzymes, 43–44
Micrococcus lysodeikticus, 32–33, 45, 259
Modulation, 96–97
Mutations, 54, 56, 101; acridine, 144; amber, 74, 76, 117–23; changes, 54; deletion, 116; ochre, 118–20; point, 102; suppressor, 61, 107, 123, 125
Mutagens, chemical, 7, 107
Myoglobin, 150, 152–54, 165, 170, 179, 185

Nearest-neighbor frequencies, 20, 30, 32, 242–43; analyses, 57, 230
Neurospora crassa, 78, 93, 128, 195, 205
Nitrosoguanidine, 122

Index

Nitrous acid, 23, 103, 108, 111–12, 114, 133, 138–42
Nonsense mutation, 115
Nonsense triplet, *see* Chain-terminating triplets

Ochre mutations, 118–20
Ochre triplet, 118–20
Operator gene, 95
Operon, 93, 95, 169–70
Origin of life, 3, 64, 245–46; amino acids, 65; purines, 65; pyrimidines, 65
Origin of Species, The, 261
Overlapping code, 23

Pantothenic acid, 260
Peptidases, 18
Phosphatase, alkaline, 60, 106, 121–25
Phosphoglucomutase, 259
Photoprotection, 84
Photoreactivation, 84
Point mutations, 102; evaluation, 131
Poisson distribution, 175, 188, 200, 205, 214, 219
Polymorphism, 185
Polypeptide synthesis, 44, 87
Polyploidy, 7
Polysomes (polyribosomes), 25, 36, 45, 86
Precipitin test, 255
"Primeval nutritive soup," 64, 187, 250
Proflavine, 110, 113
Proteus vulgaris, 38
Protoporphyrin, 149
Pseudomonas aeruginosa, 38, 193
Pseudouridine, 24
Purines, 65, 180
Pyrimidines, 65

Rapid lysis, 107
Recognition site, 24, 27, 40, 70
Ribosomes, 23–27, 37, 43, 54–55, 60–61, 69, 77, 85–89
Repression, coordinate, 96
Ribosomal RNA, 23, 37; attachment, 61; binding, 59–61, 77
Ribosomal proteins, 87
RNA, *see* Messenger RNA, Ribosomal RNA, Transfer RNA
RNA polymerase, 22, 34, 36–44, 90–92

Scopes trial, 262
Salmonella typhimurium, 93, 98, 242
Segregation-distorter gene, 127
Sickle-cell anemia, 125, 164–68, 184
sRNA, *see* Transfer RNA
Strand selection, 89–92; hybrid, 8, 35, 81
Streptomycin, 21
Structural genes, 95
Suppression, 115, 116, 118, 124; suppressor mutant, 61, 107, 123, 125; suppressor strains of *E. coli*, 74
Synthetase: animoacyl-sRNA, 77; *E. coli*, 78; leucyl-sRNA, 79; tryptophan, 54, 70, 106, 133–37.

Thymine dimers, 84
Thyroxine, 254–55
Tobacco mosaic virus, 54, 137–42, 215–18; *dahlemense* strain, 215–18; *vulgare* strain, 215–18
Transfer RNA (sRNA), 39–41; functional regions, 24, 77–80; molecules, 25, 27, 37, 77; alanyl, 70, 77–79; leucyl, 78; fractionation, 77–78; synonymous, 40; coding triplets in, 72.
Transformation, 82
Transitions, 103, 108, 110, 141–42, 215–18
Transversions, 103, 105, 110, 132–33, 145, 215–18
Trypsinogen, 127, 209–15
Tryptophan, biosynthetic pathway, 65; synthetase, 54, 70, 106, 133–37
2-Aminopurine, 107–9, 113, 114, 119, 136
Tyrosine, biosynthetic pathway, 65

Ultraviolet light, 103, 113
Ultraviolet reactivation, 84
Unusual bases, sRNA, 40–44

Vitamin A, 250–55; and metamorphosis, 250–52

Watson-Crick pairing, 3, 5, 13, 81, 91, 102, 139, 231
Wilberforce, Bishop Samuel, 261
"Wobble" hypothesis, 71

X-rays, 101, 103, 107